Fortschritte Naturstofftechnik

Reihe herausgegeben von
T. Herlitzius, Technische Universität Dresden, Deutschland

Die Publikationen dieser Reihe dokumentieren die wissenschaftlichen Arbeiten des Instituts für Naturstofftechnik, um Maschinen und Verfahren zur Versorgung der ständig wachsenden Bevölkerung der Erde mit Nahrung und Energie zu entwickeln. Ein besonderer Schwerpunkt liegt auf dem immer wichtiger werdenden Aspekt der Nachhaltigkeit sowie auf der Entwicklung und Verbesserung geschlossener Stoffkreisläufe. In Dissertationen und Konferenzberichten werden die wissenschaftlich-ingenieurmäßigen Analysen und Lösungen von der Grundlagenforschung bis zum Praxistransfer in folgenden Schwerpunkten dargestellt:

- Nachhaltige Gestaltung der Agrarproduktion
- Produktion gesunder und sicherer Lebensmittel
- Industrielle Nutzung nachwachsender Rohstoffe
- Entwicklung von Energieträgern auf Basis von Biomasse

Weitere Bände in der Reihe http://www.springer.com/series/16065

Amer Khalid Ahmed Al-Neama

Evaluation of performance of selected tillage tines regarding quality of work

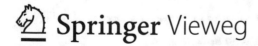

Amer Khalid Ahmed Al-Neama
Dresden Technical University
Dresden, Germany

A dissertation submitted to the Faculty of Mechanical Science and
Engineering at the Technische Universität Dresden for the degree of Doctor
of Engineering (Dr.-Ing.)
by
Amer Khalid Ahmed Al-Neama

Reviewer:
Prof. Dr.-Ing. habil. Thomas Herlitzius
Prof. Dr.-Ing. Till Meinel

Day of Submission: 21.09.2017
Day of Defense: 05.03.2018

Chairman of the Promotion Commission: Prof. Dr.-Ing. habil. Thorsten Schmidt

ISSN 2524-3365   ISSN 2524-3373   (electronic)
Fortschritte Naturstofftechnik
ISBN 978-3-662-57743-1   ISBN 978-3-662-57744-8   (eBook)
https://doi.org/10.1007/978-3-662-57744-8

Library of Congress Control Number: 2018946852

Springer Vieweg
© Springer-Verlag GmbH Germany, part of Springer Nature 2019

This Springer Vieweg imprint is published by Springer Nature, under the registered company
Springer-Verlag GmbH, DE
The registered company address is: Heidelberger Platz 3, 14197 Berlin, Germany

Dedication

This work is dedicated to my brother Mohammed Khalid Ahmed AL-Neama

Acknowledgement

I express my sincere appreciation and gratitude to my supervisor professor Thomas Herlitzius for his guidance, invaluable advice, and encouragement during these studies and during the preparation of the thesis.

Very special thanks to Mr. Tim Boegel, Mr. Andre Grosa, and Mr. Bruno Planitz, the team in the soil bin laboratory.

My thankfulness goes to my colleagues in the Chair of Agricultural Systems and Technology Department.

I would like to express my gratefulness to University of Diyala and the Ministry of Higher Education and Scientific Research in Iraq for the financial support.

Finally, I would like to thank my parents for their encouragement throughout this study

List of Symbols and Abbreviations

A	Area [cm²]
Aa	All-free area [cm²]
Af	Area of soil remaining in furrow after tillage [cm²]
Ar	Area of ridge [cm²]
At	Total area [cm²]
Aw	Area of the furrow [cm²]
AST	Chair for Agricultural Systems and Technology
C	Soil cohesion [kPa]
Ca	Soil adhesion [kPa]
$C_0, C_1, C_2, C_3, C_4, C_5$	Regression coefficients
C_I	Cone index [kPa]
CT	Conservation Tillage
d	Depth [cm]
d.b.	Dry base
dc	Critical depth [cm]
Ds	Distance [cm]
2-D	Two-dimensional
3-D	Three-dimensional
F	Force [kN]
f	Function
F – Value	Statistical value
Fc	Cohesion force [kN]
Fca	Adhesion force [kN]
Fd	Draft force [kN]
Ff	Friction force [kN]
Fg	Gravitational force [kN]
Fh	Horizontal force [kN]
Fi	Inertial force [kN]
FL	Lateral force [kN]
Fn	Normal force [kN]

Fp	Penetration force [kN]
Fs	Specific force [kN/m²]
Ft	Total force [kN]
ft.	Feet
Ft_f	Forward failure force component [kN]
Ft_s	Sideways failure force component [kN]
Fv	Vertical force [kN]
g	Acceleration of gravity [m/s²], Gram
G	Coefficient of static resistance [kN/m²]
G_1, G_2, G_3	Tine geometric regression coefficients
GLM	General linear model (statistical analysis)
HF	Height of furrow [cm]
HR	Height of the ridge [cm]
in.	Inch
K	Dummy variable regression coefficient
Ka	Critical aspect ratio
Lt	Tine length [cm]
m	Rupture distance ratio
Mc	Soil moisture content [%]
MC_d	Soil moisture content dry base [%]
MC_w	Soil moisture content wet base [%]
MWT	Maximum width of soil throw [cm]
N	Dimensionless factors
Na	Dimensionless factors denote the soil inertia
Nc	Dimensionless factors denote the cohesiveness
Nca	Dimensionless factors denote the adhesiveness
Nq	Dimensionless factors denote the surcharge pressure
Nγ	Dimensionless factors denote the gravity
n.s.	Not significant
P – Value	Statistical value
Ps	Specific power [kW/m²]
q	Surcharge pressure [kPa]

r	Rupture distance [m]
R^2	Coefficient of determination
RRD	Ridge to ridge distance [cm]
S	Speed [km/h]
SD	Standard deviation
SLa	Soil loosening percentage above the soil surface [%]
SLu	Soil loosening percentage under the soil surface [%]
SPSS	Statistical Package for Social Scientists (statistical program)
T	Tine, Treatment
T1	Heavy duty tine
T2	Double heart tine
T3	Double heart tine with wing
T4	Duck foot tine
TU Dresden	Technische Universität Dresden
USDA	United States Department of Agriculture
V	Volume [m³]
Vs	Core sample volume [cm³]
w	Width [cm]
Wd	weight of dry soil [g]
Wf	Width of furrow [cm]
Wm	weight of wet soil [g]
Wt	Tool width [cm]
Ww	Wing width [cm]
α	Rake angle [°]
β	Soil failure angle [°]
δ	Soil-metal friction angle [°]
ϕ	Soil internal friction angle [°]
λ	Extended angle of side crescent [°]
θt	Tine angle [°]
τ	Soil shear strength [kPa]
σ	Soil normal stress [kPa]
ρ	Soil bulk density [g/cm³]

ρ_w	Wet soil bulk density [g/cm³]
ρ_d	Dry soil bulk density [g/cm³]
γ	Specific weight [N/m³]
ω	Constant of dynamic draft [kN s²/m4]

Table of Contents

1 Introduction

1.1 Tillage

Soil preparation is the first and most fundamental step in crop production. Therefore, the main purpose of tillage is to prepare the growth zone for planting in the soil. Generally, tillage is bounded to the arable soil, which involves organic matter and is limited to a depth of about 10 to 90 cm. Another objective of tillage is to improve the soil's physical properties, for example weed control etc. (Stout & Cheze 1999).

1.2 Soil Erosion (Problem)

Soil erosion is action caused by wind or water forces, which affects the Earth's crust. Soil erosion is a serious driver for desertification, degrading the soil's capacity to sustain vegetation under harsh conditions. Global climate change could have significant influence on soil erosion, but this influence has received little research concern since it contains complex interactions between multiple factors dominating erosion rates. Van-Camp et al. (2004) identify different forms of soil erosion processes. As a natural process, it was caused by wind or water erosion; human activities have caused further disturbances or translocation erosions such as tillage erosion, land leveling, and harvesting of root crops.

Tillage erosion is determined by the type and design of the tillage tools and the way that tillage tools stir the soil. It is highly related to traditional tillage or conventional tillage through the use of moldboard plows and disk plows. Boardman & Poesen (2006) found that the rate of tillage erosion has increased recently in Europe due to the increases in tillage speed and depth, which have increased the rate of soil translocation.

1.3 Tillage Systems (Solution)

A new approach to tillage is needed to decrease the quantity of erosion, soil degradation, soil compaction, and high fuel consumption. These undesired effects are caused by intensive tillage, known as the traditional tillage system. This system involves using moldboard or disc plows.

Recently, many economic and environmental changes have taken place in the agricultural industry. A better understanding of the environmental impacts now requires farmers to review their tillage systems and to implement new techniques of soil cultivation. This new technique, called the conservation tillage system, is strongly related with lower energy consumption and soil erosion.

The USDA (United States Department of Agriculture) defines conservation tillage (CT) as "any tillage system that maintains at least 30 % of the soil surface covered by residue where the objective is to reduce erosion". Fig. 1.1 shows the difference between traditional tillage and conservation tillage regarding to the residue.

Conservation tillage systems have benefits other than soil conservation, such as increased water infiltration, increased or sustained organic matter content, increased water-holding capacity, and continued long-term productivity of the soil.

© Springer-Verlag GmbH Germany, part of Springer Nature 2019
A. K. A. Al-Neama, *Evaluation of performance of selected tillage tines regarding quality of work*, Fortschritte Naturstofftechnik, https://doi.org/10.1007/978-3-662-57744-8_1

Fig. 1.1 Tillage systems: A) Traditional tillage without residue on the soil surface (LEMKEN GmbH & Co. KG) and B) Conservation tillage with residue on the soil surface (AST)

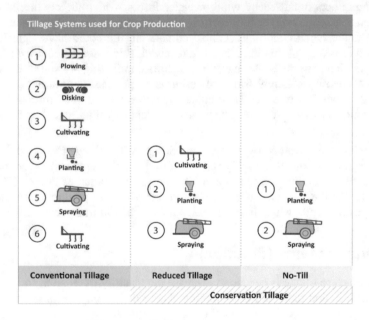

Fig. 1.2 Tillage systems used for crop production

It also requires less capital investment in equipment and fewer field passes, which reduces the amount of labor and fuel used (Baker & Saxton 2007). There are five types of tillage in conservation tillage: No-tillage or zero tillage, Mulch tillage, Strip or zonal tillage, Ridge tillage, and Reduced or minimum tillage (Fig. 1.2).

1.4 Chisel Plow

The chisel plow is considered one of the primary passive tillage tools; it is commonly used to till land with minimal soil disruption. But its function in the soil was different compared to the others. Sometimes the farmer would use the chisel plow as the secondary tillage tools.

- Chisel plowing will not invert or turn the soil, so it is often used with no-till and low-till farming practices that seek to reduce erosion.
- Reduces the effects of compaction.
- Weed control.
- Increases water infiltration.
- Can help break up hardpan.
- Aids seedbed preparation.

Besides all of those advantages, a chisel plow has a simple design: frame fixed shanks, which end with tines (see Fig. 1.3).

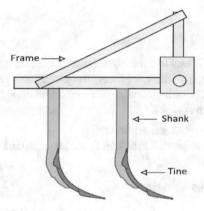

Fig. 1.3 Chisel plow

Shanks can be rigid or flexible (spring behaviors). All further results are assuming tines mounted on rigid shanks.

1.5 Tines

Soil disturbance by tines is an important part of farming operations. Tines are used in chisel plows, subsoil plows, liquid injection, seeders, spike tooth harrows, and planters. Sometimes they are attached as an additional tool for other tillage equipment and used in heavy farm machinery.

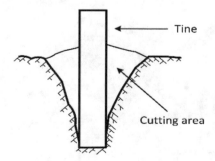

Fig 1.4 The effect of a tine on soil (Godwin & O'Dogherty 2007)

Koolen & Kuiper (1983) classify tillage tools into four categories: tines plow bodies, discs, and rollers. The soil cutting effect of tines reaches further than the width of the tine body (see Fig 1.4), while for plow bodies the soil cutting effect is generally as wide as the body width.

Godwin & O'Dogherty (2007) classified tines into three categories according to the depth/width (d/w) ratio, which includes the following:

- Wide tines (blades) where d/w < 0.5.
- Narrow tines (chisel) where 1 < d/w < 6.
- Very narrow tines (knife) where d/w > 6.

1.6 Objectives

The main aim of this work was to evaluate the performances of chisel plows by testing a varying singular tine shape relating to forces required, soil profile parameters, and its dependency on the main operating conditions (speed and depth) under controlled soil bin conditions in a sandy loam soil.

- Developed new regression equations.
- Finding a model to predict the draft force for standard single tine.

1.7 Structure of this Work

This work is divided into seven major parts. Following this introduction, Chapter 2 summarizes the state of the art in terms of standard or plain single tine. This chapter is divided into two minor parts. The first one is related to the forces and the second one is related to the soil profile. Chapter 3 illustrates the method of this work: tines, devices, equations, programs, and locations of this work. Chapter 4 explains the results obtained from this work while Chapter 5 compares selected results with other works. Chapter 6 explains a model to predict draft force by using principles of soil mechanics and soil profile evaluation. Finally, chapter 7 concludes all results.

2 Literature Review

This chapter is divided into two main sections. The first section presents background on forces acting on the single tine, a brief illustration of the most famous draft force analytical models. The second section is related to the soil profile parameter.

2.1 Basic Concept

First of all, it is important to specify the forces acting on the single tine. For simple tillage tools (tines), only two components are considered (see Fig. 2.1).

- Fh is the horizontal force required to pull or push the tine through the soil.
- Fv is the vertical force assisting or preventing penetration into the soil.
- FL is the lateral force, it is equal to the zero.
- Fd is the total force or draft force, it is the result between Fh and Fv.

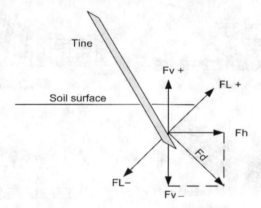

Fig. 2.1 Forces acting on the single tine

Ideally, Fh must be as small as possible, Fv must be directed downwards (-) to assist penetration for major soil loosening operations, and, if FL is equal to any amount, that means there was an adjustment error.

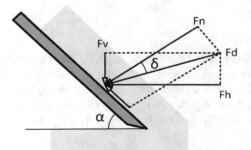

Fig. 2.2 Normal force and friction force generated by soil-tine interface (Koolen & Kuipers 1983)

© Springer-Verlag GmbH Germany, part of Springer Nature 2019
A. K. A. Al-Neama, *Evaluation of performance of selected tillage tines regarding quality of work*, Fortschritte Naturstofftechnik, https://doi.org/10.1007/978-3-662-57744-8_2

Two forces are exerted with soil-tine interface. Fn is the normal force that the soil exerts on the tool surface (perpendicularly to the tine). Ff is the friction force between the soil and the tool surface (tangentially on the tine) (see Fig. 2.2). δ and α are soil-metal friction angle and rake angle (cutting angle), respectively.

2.1.1 Disturbance

The objective of tillage is to disturb a given soil body, which is called a soil failure in literature. Several tool angles, travel speeds, and depths were used in order to extinguish the pattern of soil cutting failure in front of the tine. Koolen & Kuipers (1983) describe three patterns of soil failure by tine: the shear type, the flow type, and the type with open crack formation. Four distinct types of soil failure were identified by Elijah & Weber (1971): shear-plane, flow, bending, and tensile (see Fig. 2.3). Meanwhile, Aluko & Seig (2000) identified five types of soil failure: shear failure, plastic flow, crescent failure, lateral failure, and tensile failure.

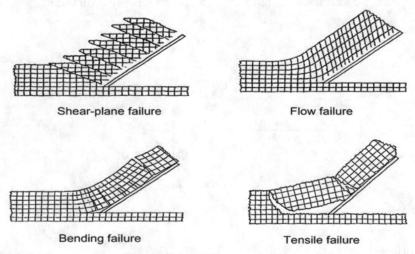

Fig. 2.3 Soil failure patterns (Elijah & Weber 1971)

Numerous research papers have focused on the shear-plane failure pattern with soil parameters as a model to predict the cutting force (Payne 1956, Söhne 1956, Osman 1964, Siemens et al. 1965, Hettiaratchi & Reece 1967, Godwin & Spoor 1977, McKyes & Ali 1977) and others.

2.1.2 Critical Depth

Generally, a tine working below the critical depth dc causes lateral failure (see Fig. 2.4). The soil moves forwards and sideways only. Spoor & Godwin (1978) reported that the critical depth depended on soil conditions, operating depth, tine width, and rake angle. The magnitude of dc was empirically found by Godwin & Spoor (1977). However, its location is not known broadly Godwin & O'Dogherty (2007).

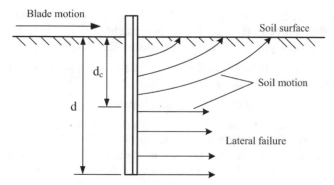

Fig. 2.4 Lateral soil failure and critical depth (McKyes 1985)

From the agricultural point of view, a tine operated below the critical depth is undesirable (Dedousis & Bartzanas 2010). It reduces soil distributions, and increases soil resistance and compaction (Kasisira 2004).

2.2 Forces

2.2.1 Coulomb's Low

The most basic developments in understanding soil failure were developed by Coulomb in 1776. He provided the first comprehensive description of soil shear strength. He assumed that the shear strength of soil consisted of two components: cohesion and friction. He identified an equation for the force per unit area acting on a surface failure in the soil body as follows.

$$\tau = c + \sigma \tan \phi \qquad (2.1)$$

Where: τ is shear strength (force per unit area), σ is normal stress, c is soil cohesion (force per unit area), and ϕ is the soil internal friction angle.

The force per unit area acting at the metal-soil interface is determined by the following equation.

$$\tau = c_a + \sigma \tan \delta \qquad (2.2)$$

Where: c_a is soil adhesion at the soil-tool interface (force per unit area) and δ is the external frictional angle at the soil-tool interface.

2.2.2 Draft Force Predicted and Calculated

Draft force prediction magnitude can be calculated by using one of these methods.

- Dimensional analysis.
- Analytical models: A) Two-dimensional and B) Three-dimensional.
- Empirical (soil bin, field experiment) regression equation.
- Numerical model such as Finite Element Analysis (FEA).

2.2.2.1 Dimensional Analysis

Using dimensional analysis requires only that all pertinent physical quantities that impact the process be identified. These are then consolidated into groups, with each group being dimensionless.

Numerous researches have used the principle of dimensional analysis to reduce the number of independent variables (Osman 1964, Hettiaratchi et al. 1966, Sprinkle et al. 1970, Luth & Wismer 1971, Upadhyaya et al. 1984, Moeenifar et al. 2013, 2014).

Sprinkle et al. (1970) indicated that the draft force of a plane blade at fixed length to width ratio is presented in Eq. (2.3).

$$Fd = f\,(S, g, d, \alpha, \phi, \delta, c, \rho) \tag{2.3}$$

Where: ρ is the soil bulk density, d is working depth, S is operating speed, and g is acceleration of gravity. These quantities include the dimensions of force, length, and time. The sum of those dimensions is three and, according to the Pi-theorem, the eight parameters can be reduced to five, as shown in Eq. (2.4).

$$\frac{Fd}{\rho \cdot d^3} = f\left(\frac{S^2}{g \cdot d^2}, \alpha, \phi, \delta, \frac{c}{\rho \cdot d}\right) \tag{2.4}$$

While Upadhyaya et al. (1984) stated that the draft force of any passive chiseling tine was found to be a function of operation conditions, tool geometry, and soil properties, as given by Eq. (2.5), they assume that the soil properties, such as texture, internal friction angle, and soil metal friction angle, are related to the cone index.

$$Fd = f\,(\rho_w, C_I, d, S, w, L, \alpha, g) \tag{2.5}$$

Where: ρ_w is the wet bulk density of the soil, C_I is the cone index, W is tool width, and L is tool length. Using the dimensional analysis procedure and Buckingham Pi theorem on Eq. (2.5), the result is Eq. (2.6).

$$\left\{\frac{Fd}{C_I\,W^2}\right\} = f\left[\left\{\frac{\rho_w\,S^2}{C_I}\right\}, \left\{\frac{d}{w}\right\}\right] \tag{2.6}$$

The main problem of this method is that it develops a solution for a specific condition rather than a general solution. Moreover, the Pi-terms may have the difficulty of being too complex to understanding.

2.2.2.2 Analytical Models

All analytical models are based on Terzaghi's passive earth pressure theory (Terzaghi 1943). According to Terzaghi's theory, a failure zone is assumed to exist ahead of a cutting blade and the soil in the failure zone is assumed to be in the critical failure state. The resulting force on the blade was calculated by assuming static equilibrium along the boundary of soil failure. The limit equilibrium approach can be used to achieve only the information about the maximum forces which generated inside the soil body without providing many clues about how the soil body deforms (Shen & Kushwaha 1998).

2.2.2.2.1 Two-Dimensional Analytical Model

The predictive soil cutting force model was first developed for a two-dimensional soil cutting for a wide blade aspect ratio (depth to width ratio less than 0.5). The soil moved in the upward and forward directions only. Reece (1965) proposed a universal earthmoving equation (UEE) for a wide blade at low speed as shown in Eq. (2.7).

$$Ft = (\gamma\, d^2\, N_\gamma + cd\, N_c + c_a d\, N_{ca} + qd\, N_q)\, w \tag{2.7}$$

Where: Ft is the soil resistance force or soil cutting force or total force, γ is the total specific weight of the soil, q is the surcharge pressure vertically acting on the soil surface, and N_γ, N_c, N_{ca}, and N_q are dimensionless factors denoting the gravitational, cohesive, adhesive, and surcharge components of the soil reaction, respectively. These *N factors* are functions of the angle of internal fraction, the angle of the soil-metal friction, and the rake angle. Hettiaratchi et al. (1966) revised Reece's model and presented a set of charts with calculated values of *N factors*. Sokolovski (1965) used a numerical solution for these *N factors*. The magnitude of soil-metal adhesion was very small. Therefore, Hettiaratchi & Reece (1974) developed Eq. (2.7) by combining the cohesive and adhesive terms as presented in Eq. (2.8).

$$Ft = (\gamma \, d^2 \, N_\gamma + cd \, N_c + qd \, N_q) \, w \qquad (2.8)$$

2.2.2.2.2   Three-Dimensional Analytical Models

In practice, tillage is mostly carried out with the use of a narrow tine (depth to width ratio between 1 and 6). Therefore, semi empirical three-dimensional models have been developed for narrow and very narrow tines (depth to width ratio greater than 6).

There are two distinguished failure zones related with narrow and very narrow tines (see Fig. 2.5) operating above the dc. Zone 1 in front of the tine is similar to the two-dimensional failure zone of wide tine, zones 2 on either side of zone 1 are the side crescents of soil failure.

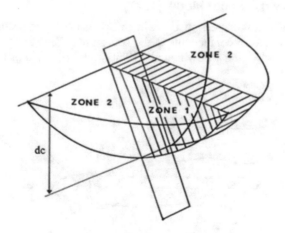

Fig. 2.5 Three-dimensional soil failure zones (Desir, 1981)

The first three-dimensional analytical model for an inclined narrow tine was developed by Payne (1956). Hettiaratchi & Reece (1967) developed a three-dimensional model for soil failure based on the Reece's universal earth-moving equation.

O'Callaghan & Farrelly (1964) proposed a model that included a critical depth, as well as further developing Payne's model, for which they assumed that the critical depth d_c is equal to 0.6 times the tine width. Godwin & Spoor (1977) modified Eq. (2.8) for a narrow tine by taking into consideration the crescent shaped failure patterns. They also proposed a model to calculate the lateral force component under the critical depth. This prediction model was developed further by Godwin et al. (1984). McKyes & Ali (1977) proposed a soil failure wedge consisting of a center wedge and two circular side crescents ahead of a tool, similar to the Godwin & Spoor soil failure pattern (but with a different shape). Perumpral et al. (1983) proposed another three-dimensional model for a narrow tine. The model assumed two forces (two side crescents) acting on the side of the center wedge (similar to the McKyes & Ali soil failure form). The total force was determined by considering the equilibrium of all forces acting on the soil failure perimeter.

2.2.2.2.2.1 Hettiaratchi and Reece Model (1967)

The prediction equation for Hettiaratchi & Reece's model is presented in Eq. (2.9) and (2.10).

$$Ft_f = (\gamma \, d^2 \, N_\gamma + cd \, N_c + c_a d \, N_{ca} + qd \, N_q) \, w \tag{2.9}$$

$$Ft_s = \left[\gamma (\, d + q/\gamma)^2 \, w \, N_{s\gamma} + cwd \, N_{sc} + qd \, N_q\right] K_a \tag{2.10}$$

Where: Ft_f is forward failure force component, Ft_s is sideways failure force component, and K_a is the critical aspect ratio. The horizontal and vertical components of the total force can be given in Eq. (2.11) and (2.12), respectively.

$$Fh = Ft_f \, Sin \, (\alpha + \delta) + Ft_s \, Sin \, \alpha + c_a d \cos \alpha \tag{2.11}$$

$$Fv = Ft_f \, cos \, (\alpha + \delta) + Ft_s \, cos \, \alpha + c_a d \tag{2.12}$$

2.2.2.2.2.2 Godwin & Spoor Model (1977)

The soil failure pattern proposed by Godwin & Spoor (1977) is presented in Fig. 2.6. The horizontal and vertical components of the soil cutting force are given by Eq. (2.13) and (2.14), respectively, after Godwin et al. (1984) and further simplified by Godwin & O'Dogherty (2007).

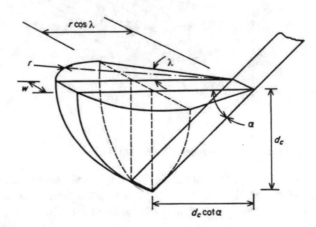

Fig. 2.6 Soil failure zone of Godwin & Spoor (1977)

$$Fh = \left(\gamma \, d_c^2 N_\gamma + cd_c \, N_c + qd_c \, N_q\right)$$
$$\times \left[w + d\left\{m - \frac{1}{3} \, (m - 1)\right\}\right] \sin(\alpha + \delta) \tag{2.13}$$

$$Fv = -\left(\gamma \, d_c^2 \, N_\gamma + cd_c \, N_c + qd_c \, N_q\right)$$
$$\times \left[w + d\left\{m - \frac{1}{3} \, (m - 1)\right\}\right] \cos(\alpha + \delta) \tag{2.14}$$

Where: m is the rupture distance ratio (the ratio between forward rupture distance r and working depth d). The term $d\{m-\frac{1}{3}(m-1)\}$ refers to the side effect of the tine on the crescent failure.

2.2.2.2.2.3 McKyes and Ali Model (1977)

Figure 2.7 illustrates the soil failure zone of the McKyes and Ali model.

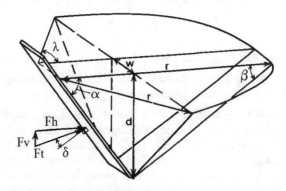

Fig. 2.7 Soil failure zone of McKyes & Ali (1977)

The prediction equation for the horizontal force is presented in Eq. (2.15). However, the N-factors have been reassessed for three dimensions, which is given as

$$Fh = (\gamma d^2 N_{\gamma H} + cd\, N_{cH} + qd\, N_{qH})\, w \tag{2.15}$$

$$N_{\gamma H} = \frac{\frac{r}{2d}\left[1 + \frac{2r}{3w}\sin\lambda\right]}{\cot(\alpha + \delta) + \cot(\beta + \phi)} \tag{2.16}$$

$$N_{cH} = \frac{[1 + \cot\beta\,\cos(\beta + \phi)]\left[1 + \frac{r}{d}\sin\lambda\right]}{\cot(\alpha + \delta) + \cot(\beta + \phi)} \tag{2.17}$$

$$N_{qH} = \frac{\frac{r}{d}\left[1 + \frac{r}{d}\sin\lambda\right]}{\cot(\alpha + \delta) + \cot(\beta + \phi)} \tag{2.18}$$

Where: λ is extended angle of side crescent (see Fig. 2.7), and r is forward rupture distance of the failure plane on the surface, which is given by Eq. (2.19).

$$r = d(\cot\alpha + \cot\beta)\, w \tag{2.19}$$

Where: β is the soil failure angle.

McKyes & Desir (1984) developed the model by adding the effect of adhesion. The total cutting force can be written as

$$Ft = (\gamma d^2 N_{\gamma H} + cd\, N_{cH} + qd\, N_{qH})\, \frac{w}{Sin(\alpha + \delta)}$$
$$+ \frac{c_a d\,[1 - \cot\alpha\,\cot(\beta + \phi)]\, w}{\cos(\alpha + \delta) + \sin(\alpha + \delta)\cot(\beta + \phi)} \tag{2.20}$$

The horizontal component of the force Ft can be formed as

$$Fh \; = \; Ft \sin(\alpha + \delta) + c_a dw \cot \alpha \tag{2.21}$$

Or

$$Fh = \left(\gamma \, d^2 N_{\gamma H} + cd \, N_{cH} + qd \, N_{qH} + c_a d \, N_{aH} \right) \tag{2.22}$$

And

$$N_{aH} \; = \; \frac{1 + \cot \alpha \cot(\alpha + \delta)}{\cot(\alpha + \delta) + \cot(\beta + \phi)} \tag{2.23}$$

While the vertical component of the force Ft given by McKyes (1985) as

$$Fv \; = \; Ft \cos(\alpha + \delta) - c_a dw \tag{2.24}$$

2.2.2.2.3   Dynamic Analytical Model

All models mentioned above are static models without considering the effect of travel speed, which is common to tillage practice.

Stafford (1979, 1984) proposed dynamic models based on Hettiarratchi and Reece's static models for both two and three-dimensional soil failure cases by introducing acceleration effects into these models. McKyes (1985) added an inertia term to the universal earth-moving equation for a wide blade. Swick & Perumpral (1988) proposed a three-dimensional dynamic soil failure model; the proposed soil failure zone is similar to McKyes & Ali's static model and the force equation is derived in the same way except an acceleration force is included to account for the travel speed effect. Zeng & Yao (1992) also developed a dynamic model. This was obtained from the relation between soil shear strength and shear strain rate. The failure zone is also similar to McKyes & Ali's model. Onwualu & Watts (1998) improved McKyes & Ali's model with a dynamic for a narrow tine.

Wheeler & Godwin (1996) empirically developed (Godwin et al. 1984) a static model for single and multi-narrow tines operating at speeds until 20 km/h. They also formed an additional equation that considers both the number of tines and the spacing between them.

The general form of a dynamic model is presented in Eq. (2.25), as given by McKyes (1985).

$$Ft \; = \; (\gamma \, d^2 \, N_\gamma + cd \, N_c + c_a d \, N_{ca} + qd \, N_q + \rho \, S^2 d \, N_a) \, w \tag{2.25}$$

Where: N_a is a dimensionless factor for soil inertia effects, is a function of α, δ, ϕ and β, and is given by Eq. (2.26). Other notations as mentioned previously.

$$N_a \; = \; \frac{\tan \beta + \cot(\beta + \phi)}{(\cos(\alpha + \delta) + \sin(\alpha + \delta) \cot(\beta + \phi)) \, (1 + \tan\beta \, \cot\alpha)} \tag{2.26}$$

2.2.2.2.4   Summary of Analytical Approach

Below is a summary of the analytical models.

- All models were based on Terzaghi's passive earth pressure theory and empirical work.
- A plane blade with known inclination angle was used.
- A shear-plane soil failure pattern was proposed in front of blade.

- Soil moved horizontally and vertically only (center of wedge) for a wide blade and moved also laterally (side crescents of wedge center) for narrow and very narrow tines.
- The two-dimensional model is valid only for a wide tine.
- The three-dimensional model is valid only for a narrow tine.
- Soil cutting force was calculated based on a static equilibrium along the perimeter of the soil failure zone.

As mentioned above, all these models used a simplified flat blade with known rake angle, neglecting standard tine shapes, which are curved or with wings. Therefore, these models have limitations in evaluating tillage tines.

2.2.2.3 Empirical Regression Equation

Empirical models for predicting draft force using statistical regression equations based on data collected from field experiments or soil bin for various tillage tools at various soil and operating conditions were adopted (Upadhyaya et al. 1984, Grisso et al. 1996, Onwualu & Watts 1998, Rosa & Wulfsohn 2008, i.a.). However, those regression equations are limited to the tillage tools and soil conditions tested. New regression equations using reference tillage tools have been developed (Glancey & Upadhyaya 1995, Glancey et al. 1996, Desbiolles et al. 1997, Ehrhardt et al. 2001, Sahu & Raheman 2006, i.a.). All these regression models calculate the draft force as a ratio between the test sample and reference model without considering effects of the tool's geometry on draft force.

2.2.2.3.1   Effect of Operation Conditions

Increasing speed means that a tine moves faster through the soil; generally, the draft force increases with speed, which is often attributed to the acceleration of soil (Dedousis & Bartzanas 2010). Schuring & Emori (1964) reported that a speed below $\sqrt{5gw}$ was not significant for draft force, which equals 4.4 km/h. Wheeler & Godwin (1996) proved Schuring & Emori's equation for a single and multi-narrow tine and they also stated that the speed becomes critical at $\sqrt{5g(w + 0.6\,d)}$, which equals 10.7 km/h.

Many papers found a linear or second order polynomial, parabolic, or exponential relationship between the draft force and the speed (Rowe & Barnes 1961, Siemens et al. 1965, Luth & Wismer 1971, Stafford 1979, Stafford 1981, Godwin et al. 1984, McKyes 1985, Swick & Perumpral 1988, Gupta et al. 1989, Wheeler & Godwin 1996, Onwualu & Watts 1998, Sahu & Raheman 2006, Rosa & Wulfsohn 2008, i.a.). These differences in the results are attributed to the variety in soil properties and operation conditions.

Increasing depth means that a tine cuts more soil, with the draft force usually increasing dramatically with depth, which is mostly assigned to the amount of soil distributed. Furthermore, in non-cohesive soil the draft force increases linearly with depth, while in cohesive soil it increases quadratically (Koolen & Kuipers 1983). Desbiolles et al. (1997) described a quadratic behavior between the draft forces and working depth for a single standard tine in clay soil, while Sahu & Raheman (2006) described a linear behavior between the draft force and working depth for a single standard tine in a sandy clay loam soil. However, the effects of speed were found to be smaller than the effect of depth on draft force (Glancey et al. 1996, Rahman & Chen 2001, Sahu & Raheman, 2006). Therefore, Godwin (2007) recommends never working deeper than necessary.

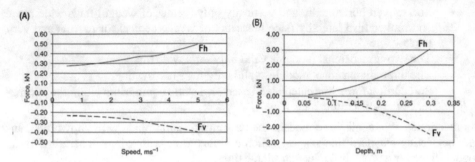

Fig. 2.8 Effect of speed and depth on horizontal (solid line) and vertical (dotted line) force: at rake
angle 30° and tine width 50 mm (Dedousis & Bartzanas 2010)

The effect of operation conditions on vertical force is illustrated in Fig. 2.8. It indicates
that the vertical force increased with speed and depth in a similar pattern but with a
smaller magnitude compared to the horizontal force.

2.2.2.3.2   Effect of Tine Geometry-Rake Angle

Several studies have extensively focused on the effects of rake angle on horizontal and
vertical forces (Payne & Tanner 1959, Dransfield et al. 1964, Godwin & Spoor 1977,
Stafford 1979, Stafford 1984, Makanga et al. 1996, Mckyes & Maswaure 1997, Onwualu & Watts 1998, Aluko & Seig 2000).

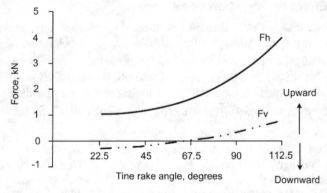

Fig. 2.9 Effect of tine rake angle on horizontal (solid) and vertical (broken) forces. Godwin &
Spoor (1977)

The effect of tine rake angle on horizontal and vertical forces is presented in Fig. 2.9,
cited from Godwin (2007). It clearly shows how both the horizontal and vertical forces
increase with rake angle.

The critical rake angle where the vertical force value changed its direction from upward
to downward is approximately 67.5°. Logically there is no vertical force at an angle
equal to 90°- δ, where the angle of soil-metal friction δ is typically approximately 22.5°.

2.2.2.3.3   Effect of Tine Geometry-Tine Width

One of the important parameters of tine geometry is the tine width. Increasing tine width
means that the furrow area increases. Generally, the draft force increases with tine width,

which is frequently referred to as the volume of soil distributed. However, the draft force increases linearly with tine width (Koolen & Kuipers 1983, Koolen 1977). Godwin (2007) demonstrated the result obtain from Godwin & Spoor (1977) showing that there is a curvilinear relationship between tine width and draft force within the range of very narrow tine width, while a linear relationship was found within the range of narrow tine width. However, the vertical force increased linearly with increasing tine width. This was confirmed by Grisso & Perumpral (1985) for a narrow tine.

Desir (1981) noticed that the draft force increases linearly with increasing tine width in a clay and sandy clay loam soil. However, the tine width has a higher significant effect on draft force in the first soil than the second soil. He also noticed a strong interaction influence between tine width and depth on draft force. Therefore, numerous studies have focused on the effect of width and depth combinations on draft force rather than individually (McKyes & Ali 1977, McKyes & Desir 1984, Mckyes & Maswaure 1997).

2.2.2.4 Comparison between Methods

Below is a comparison between methods.

- All analytical models used mathematical and physical theories, while all empirical models using statistical regression equations.
- Most of analytical models used a flat blade, while most of empirical models used standard tine.
- The empirical models took more time and consumed more money than the analytical models.

2.3 Soil profile

2.3.1 Measuring Soil Profile

The soil profile after tillage is a very important factor; it indicates and shows the result of force applied by tillage tines, which provides knowledge about the soil movement and desired disturbance (see Fig. 2.10).

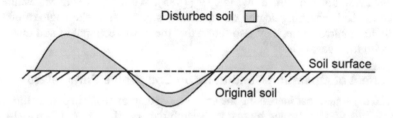

Fig. 2.10 Soil profile as cross section

Various measuring techniques have been used by researchers in order to evaluate soil profile parameters. Profile meters have developed from passing through mechanical devices such as chain rollers and pin meters, to radar scanners and optical devices like lasers and camera scanners (Willatt & Willis 1965, Saleh 1993, Mckyes & Maswaure 1997, Oelze et al. 2003, Riegler et al. 2014, Martinez-Agirre et al. 2016). Generally, profile meters can be divided into different categories: two-dimensional (2D) and three-dimensional (3D), with contact (chain roller and pin meter) and without contact (laser and camera scanner). Jester & Klik (2005) compared different methods to measure soil surface roughness, including contact methods and methods using non-contact devices, and they proved that the low cost and simple devices using contact methods like the pin meter required the highest measurement time compared to other methods, while the laser scanner provides high resolution and precise measurements. However, the laser scanner is widely affected by other sources of light (Huang & Bradford 1992, Darboux & Huang 2003). Therefore, it is preferred in lab conditions.

2.3.2 Factors Affecting Soil Profile

Liu & Kushwaha (2005) pointed out that studying the soil profile is very complex and progresses slowly because of many factors involved with it. Solhjou (2013) stated that three main factors which affect the magnitude of soil profile parameters during tillage are tool geometry (width and rake angle), operation conditions (working speed and depth), and soil properties (e.g. soil bulk density, soil moisture, etc.).

2.3.2.1 Effect of Tool Geometry on Soil Profile Parameters

Many studies mentioned that different tool geometries can create different soil profiles and soil disturbances (Sharifat & Kushwaha 2000, Chaudhuri 2001, Godwin 2007, Manuwa 2009, Manuwa et al. 2012). The tool geometry parameters for a simple plane blade can be limited to two factors: 1) tine rake angle and 2) tine width. Increasing tine width means that the width and furrow area increases, which leads to increasing the magnitude of the other soil profile parameters such as the area of the ridge and the area of soil distribution. Solhjou et al. (2013) showed that different blade face geometry caused large differences in furrow size. Willatt & Willis (1965) stated that the width and area of the furrow increased with increasing tine width for curved and plane tines. Manuwa et al. (2012) studied the effects of different tine widths on soil profile parameters under soil bin conditions and they reported that the parameters of soil disturbance except height of ridge increased with an increase in tine width. The loosened soil percentage remaining in the furrow or furrow backfill percentage increased by reducing the width of the tine (McKyes & Desir 1984, McKyes 1985). Manuwa & Ademosum (2007) noticed that the volume of soil moving in front of a tool increased with an increase in rake angle, a result also obtained by McKyes (1985). While Mckyes & Maswaure (1997) found that the soil loosening degree was generally smaller at a rake angle of 60° than at 30° or 90° for a narrow tine, they also found that the cross sectional of soil disturbed did not vary with the rake angle.

2.3.2.2 Effect of Operation Conditions on Soil Profile Parameters

Solhjou (2013) stated that increasing the speed caused an increase in soil disturbance and soil throw. This was confirmed by many research papers (Hasimu & Chen 2014, Shinde et al. 2011, Manuwa 2009, Rosa & Wulfsohn 2008, i.a.). Liu & Kushwaha (2005) reported that furrow width and the width of soil disturbance increased with increasing tillage speed, but the ridge height decreased slightly for a single sweep. Shinde et al. (2011) showed that the ridge width and furrow height increased with increasing speed and depth, and the ridge area was directly proportional to the operating speed and depth, while the maximum soil throw was obtained at the maximum speed. Hasimu & Chen (2014) observed similar behaviors for a different seed opener. McKyes (1985) reported that there is quadratic relationship between furrow area and depth for a narrow tine, while it is linear for a wide tine. Willatt & Willis (1965) found that the furrow width increased linearly with depth, while the furrow area increased quadratically with depth for a plane and curved tine at varying speeds. Mckyes & Maswaure (1997) found that the degree of soil loosening was typically superior at a large aspect ratio. Rahman et al. (2005) stated that the backfill percentage increased with operating depth and decreased with forward speed for sweep liquid injection tools.

2.3.2.3 Effect of Soil Properties on Soil Profile Parameters

Several factors related to the soil properties affect soil profile, e.g. soil moisture content, soil compaction, soil type, soil cohesion, soil adhesion, residue cover, etc. Below is a brief explanation of these factors.

Soil roughness is highly affected by the soil moisture content during tilling. The ridge height increases with increasing soil moisture content (Manuwa & Ademosum 2007). Some researchers found a negative relationship between soil moisture content and soil throw (Rosa & Wulfsohn 2008, Manuwa & Ademosum 2007, Rahman et al. 2005). Manuwa et al. (2012) found that the soil disturbance parameters decreased in value for all the tines (varying between 10 to 200 mm) as the moisture content increased from 6.0% to 17.5% (d.b.), except for height of ridge. Manuwa & Ademosum (2007) found that an increasing soil compaction leads to an increase in soil lateral throw and a reduction in ridge height. Liu et al. (2007) studied the effect of straw length on soil bin and field conditions by using sweep tools with 325 mm width. They also found that increasing straw length reduces the magnitude of soil displacement.

2.3.3 Modeling to Predict Soil Profile Parameters

Some researchers focused on developing mathematical models of the soil profile in order to predict its parameters.

Willatt & Willis (1965) performed an equation to predict the width and area of furrow for curved and plain tines as presented by Eq. (2.27) and (2.28), respectively. The furrow disturbed by both tines were roughly of a trapezoidal shape.

$$Wf = 2.42 \times d + Wt \tag{2.27}$$

Where: Wf is the width of furrow and Wt is the tine width.

$$Aw = 1.03 \times d^2 + Wt \times d \tag{2.28}$$

Where: Aw is the furrow area. Note that the measurement is in inches.

McKyes & Desir (1984) proposed a mathematical model to predict the area of the furrow as given in Eq. (2.29). For more details about notations, see Fig. 2.7.

$$Aw = d(w + r\sin\lambda) \tag{2.29}$$

Rahman & Chen (2001) found that the trapezoidal furrow shape was created by two types of sweep tools; the bottom of the trapezoid was close to the sweep width and the height to the working depth. Elsewhere the furrow was a triangular shape created by two types of disc tools with the height equal to the working depth.

Other researchers have studied soil-tool interaction with empirical models. In this approach, experimental results are modeled by regression and statistical analysis (Manuwa & Ogunlami 2010, Manuwa 2009, Manuwa & Ademosum 2007, Rahman et al.2005, Stafford 1979).

Liu & Kushwaha (2005) developed an analytical model of soil profile parameters with a single wide sweep opener in a sandy loam soil. This model can predict depth of furrow, height of ridge, and soil disturbance. However, this model can only provide soil profile parameters in the same test environments and it needs further improvement (Solhjou 2013).

2.3.4 Specific Force

From the point of designing and manufacturing, it is very important to reduce the amount of draft force acting on tillage tools with a higher quantity of soil distribution. Therefore, the specific force (force per unit area, Fs) is more useful than only force to evaluate the efficiency of tillage tools in soil cultivation (Dedousis & Bartzanas 2010). It shows the soil-tool interactions, where the force is input value and the area is output value. It has been used by several researchers (Stafford 1979, McKyes & Desir 1984, Hettiaratchi 1993, Conte et al. 2011).

Stafford (1979) stated that increasing tool speed caused increasing specific force. Although the draft force of the tine increases dramatically with depth (see section 2.2.2.3.1), this may not cause a significant increase in specific draft. Meanwhile, the cutting area increases quadratically with depth. Desir (1981) found that the specific force of a tine increases with an increase in aspect ratio, with a similar finding obtained by McKyes (1989).

Spoor & Godwin (1978) added wings to the tine to increase the effect of the tine on the soil distribution; they noticed that the area of soil distribution doubled while the draft force increased by 30 %. Therefore, the specific force can be reduced.

Mckyes & Maswaure (1997) recommended that the best tillage tool design for a low draft, high cutting percentage, and superior soil loosening should have a rake angle of about 30° and should be narrow with an aspect ratio ≥ 2.

3 Experiment Evaluation

The experiment was conducted at the indoor soil bin laboratory conditions at the Chair of Agricultural Systems and Technology (AST), Faculty of Mechanical Engineering, at the Technische Universität Dresden (TU Dresden), Germany.

In this study, three standard factors were chosen for the control of soil fitting. These standards were 1) constant soil texture, 2) constant moisture content and 3) constant soil compaction, which was represented by soil bulk density over the working depths and speeds for each tine.

3.1 Soil Bin Description and Soil Preparation

The indoor soil bin dimensions were 28.6 m long, 2.5 m wide and 1.0 m deep (see Fig. 3.1). It was filled with a sandy loam soil for which the physical properties were provided in Table 3.1.

Fig. 3.1 Soil bin laboratory

Table 3.1 Physical properties of the soil bin ± standard error

Parameters	Abbreviations	Units	Values
Soil type			Sandy Loam
Clay content		[%]	9.0
Silt content		[%]	30.1
Sand content		[%]	60.9
Dry bulk density	ρ_d	[g/cm³]	1.37 ± 0.01
Moisture content dry base	MC_d	[%]	10.4 ± 0.88
Internal friction angle	ϕ	[°]	42.0
External friction angle	δ	[°]	22.5
Cohesion	C	[kPa]	5.6

© Springer-Verlag GmbH Germany, part of Springer Nature 2019
A. K. A. Al-Neama, *Evaluation of performance of selected tillage tines regarding quality of work*, Fortschritte Naturstofftechnik, https://doi.org/10.1007/978-3-662-57744-8_3

The carriage was powered by an electric-hydraulic drive train with a maximum speed of 17 km/h delivering maximum traction of 13 kN, and was equipped with a radar sensor to measure the ground speed (velocity range 0.53 to 107 km/h ± 5%). The hydraulic drive had transversely movable three-point linkages to attach the tools and two category front and rear PTO with maximum speed 1700 rpm and maximum torque 500 Nm. Additionally, the unit had a humidification system up to 8 m³/h.

Before starting the test, the soil bin was prepared to achieve the required levels of soil moisture content and soil bulk density. The first step was plowing the soil by rotary plow at a depth of 20 cm, which was used to loosen the soil after watering to achieve the required moisture content. Following this process, the soil was leveled with the scraper and finally it was compacted by a 900 kg roller five times to obtain the required bulk density (see Fig. 3.2).

At the end of each soil preparation, samples were taken for verification of moisture content and bulk density manually by using core sample at 15 locations evenly distributed along the soil bin (see Fig. 3.3).

Fig. 3.2 Soil bin preparation

The soil moisture content and bulk density was calculated by using Eq. (3.1) and (3.2), respectively, after putting the samples in the electric oven set at 105 C° for 24 hours.

$$MC_d = \frac{(W_m - W_d)}{W_d} \times 100 \tag{3.1}$$

Where: MC_d is the soil moisture content dry base in %, W_m is the weight of wet soil in g, and W_d is the weight of dry soil in g.

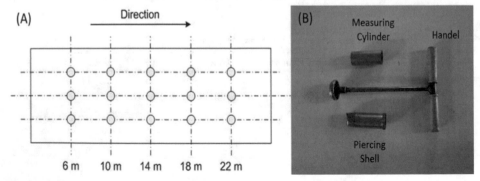

Fig. 3.3 Soil sample A) locations, B) manual core sample

$$\rho_d = \frac{W_d}{V_s} \tag{3.2}$$

Where: ρ_d is the dry soil bulk density in g/cm³ and V_s is the soil volume or measuring cylinder volume, which was 101.7 cm³.

An additional measurement was the soil penetration resistance (cone index), which was measured by using a soil penetrometer with a cone angle of 30° and cone diameter of 11.3 mm (see Fig. 3.4).

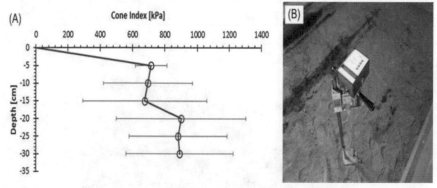

Fig. 3.4 Soil bin cone index A) mean ± standard deviation, B) soil penetrometer device SP1000

3.2 Tillage Tines Used and Measurements

Four standard chisel plow tines were used in the experiment (see Fig. 3.5): T1 Heavy Duty, T2 Double Heart, T3 Double Heart with Wings and T4 Duck Foot. Table 3.2 summarizes the characteristic parameters of the tines.

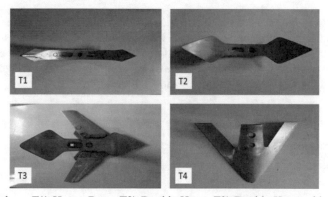

Fig. 3.5 Tines shape T1) Heavy Duty, T2) Double Heart, T3) Double Heart with wings and T4) Duck Foot

The measurements were done in two parts. Part one includes the force measurements of Fh and Fv, the horizontal and vertical forces, respectively, by using six load cell sensors, Fh (2), Fv (1), and lateral force FL (3), similar to measurements from Reich & Hohenheim (1977) (see Fig. 3.6). Sensor types were S9 and U9B (HBM GmbH) with maximum loads of 50 kN and 20 kN, respectively, with accuracy ± 5%.

Table 3.2 Tine Parameters

Tines	Length [cm]	Width [cm]	Thickness [cm]	Radius [cm]	Tine angle [°]	Weight [g]
Heavy Duty	47	6.5	2	30	60	3400
Double Heart	44	13.5	2	30	65	3200
Double Heart with Wings	32	45.0	2	30	65	4200 [1)]
Duck Foot	30	40.0	1	30	85	2900

[1)] Wing only

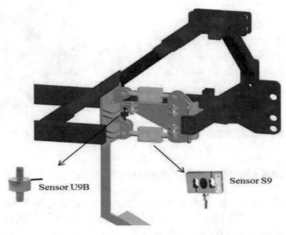

Fig. 3.6 Load cell sensors (Guhr, 2015)

Before the test run, the desired depth and speed were set up for every tine. Test data was collected in an Excel sheet and the results were plotted as diagrams as showed in Fig. 3.7. Average value and standard deviation for Fh and Fv for every tine at every depth and speed was calculated for the static range of the test, which is shown between the two vertical dashed lines in Fig. 3.7. Note that the FL is equal to zero.

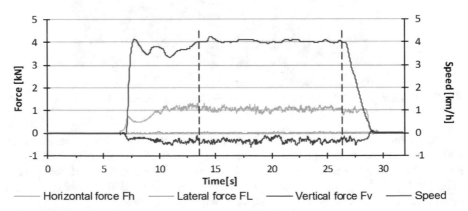

Fig. 3.7 The data obtained from the load cell sensors for the forces and from the radar sensor for the speed

Part two was related to a 2-D soil profile measurement using a laser spot sensor for the vertical coordinates and a draw wire position sensor for the horizontal coordinates (see Fig. 3.8).

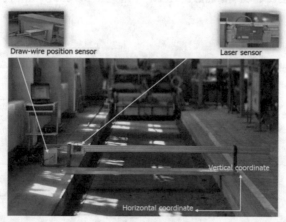

Fig. 3.8 Soil profile meter with optoNCDT 1700 from Micro-Epsilon (y-axis) and WS17KT from ASM (x-axis)

The final soil profile was acquired in three steps: first, measured the soil surface before running the test; second, measured the soil modification immediately after running the test with 21 profile readings (1 meter long every 5 cm); and third, acquired the profile after removing the loose soil. Fig. 3.9 shows the combined results of the three steps.

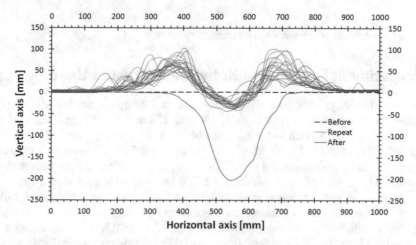

Fig. 3.9 The final soil profile for instance T1 at speed 4 km/h and depth 20 cm

3.3 Soil Profile Parameters after Tillage

For the purpose of analysis and study of soil tillage results, the soil profile after test run was divided into two parts as defined in Fig. 3.10 (a) and (b). The soil displaced above the original soil surface was called ridge and below it was called furrow (Manuwa & Ogunlami, 2010). The parameters of the soil tillage include maximum width of soil

throw MWT, ridge to ridge distance RRD, the height of the ridge HR, the height of furrow HF, the width of furrow Wf (also known as width of cut), and the tool width Wt.
Parameters of the area are area of ridge Ar, the area of soil remaining in furrow after
tillage Af, the all-free area *Aa,* and the area of the furrow cut by tines *Aw* with *Aw* equal
to Aa + Af.

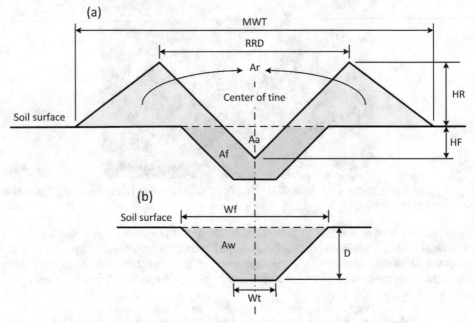

Fig. 3.10 Cross section of soil profile parameters

3.4 Experiment Design and Software Programs Used

A full factorial design was used within each tine as a complete random design (CRD).
The tines were operated at speeds 4, 7, 10, and 13 km/h for T1 and T2, 4, 8.5, and
13 km/h for T3 and T4, with varying depths of 5, 10, 15, and 20 cm for T1 and T2, and
10, 15 and 20 cm for T3. The depth of 5 cm was excluded for T3 because it is below the
minimum operational depth of the tine. T4 was run at 5, 12.5, and 20 cm depths. All tests
were done with three replicates. MATLAB based tool was used to compute the soil profile parameters. A multi-linear regression model with stepwise selection at significance
level of 5% was adopted to evaluate the significant effects of independent variables by
using SPSS statistical computer program version 22. The ANOVA table and F-value
were obtained by using GLM general linear model multivariate statistical analysis.

3.5 The Data Computations and Calculations

3.5.1 Regression Model

Three different equations of a regression model were used to predict the dependent variable for a single tine. The first model bases on the relationship between speed, depth, and

their interaction (operating conditions) and is presented by the Eq. (3.3) (Glancey & Upadhyaya, 1995).

$$Y = C_0 + C_1 S + C_2 D + C_3 SD + C_4 S^2 + C_5 D^2 \qquad (3.3)$$

Where Y is the dependent variable, S is speed in km/h, D is depth in cm, and C_0, C_1, C_2, C_3, C_4, and C_5 are the regression coefficients.

The second model is based on using a Dummy Variable K, presented by Eq. (3.4) (Al-Neama & Herlitzius, 2016). This variable represents each tine regardless of the tine shape or geometry (see Fig. 3.11).

$$Y = K + C_0 + C_1 S + C_2 D + C_3 SD + C_4 S^2 + C_5 D^2 \qquad (3.4)$$

The third model based on tines geometry is presented by Eq. (3.5) (Al-Neama & Herlitzius, 2016) (See Fig. 3.12).

$$Y = G_1 Wt + G_2 Lt + G_3 \theta t + C_0 + C_1 S + C_2 D + C_3 SD + C_4 S^2 + C_5 D^2 \qquad (3.5)$$

Where: Wt is tine width in cm, Lt is tine length in cm, θt is tine angle in ° and G_1, G_2, and G_3 are tine geometric coefficients, respectively (see Fig. 3.13). Note that identical geometric parameters were excluded (see Table 3.2).

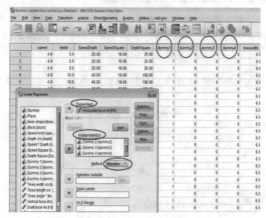

Fig. 3.11 Screen shot from Dummy Variable analysis using statistical program SPSS

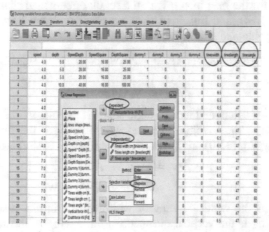

Fig. 3.12 Tines geometry analysis

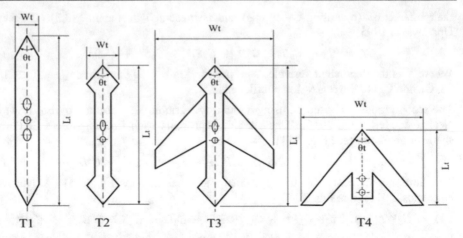

Fig. 3.13 Tine shape and dimension, where Wt is tine width, Lt is tine length, and θt tine angle

3.5.2 Calculated Soil Profile Parameters

A MATLAB based computer program was used in order to compute the distance and the area of soil profile parameters by using the following Eq. (3.6) and (3.7), respectively.

$$Ds = \sqrt{(X_2 - X_1)^2 + (Y_2 - Y_1)^2} \qquad (3.6)$$

Where: Ds is the distance between two points in cm, X and Y are the point coordinates.

$$A = \sum_{i=1}^{n} f(x_i)\,\Delta x \qquad (3.7)$$

Where: A is the area in cm².

$$\Delta x = \frac{(X_2 - X_1)}{n} \qquad (3.8)$$

Where: $x_1 \le x_i \le x_2$ is the interval, n is subinterval.

To specify the furrow shape obtained from the soil bin, the laser scanner data for every tine was compared for the best match with the regular geometric shape of a triangle or a trapezoid (see Fig. 3.14), and Fig. 3.15 shows a prediction Aw based on the geometric shape of the furrow obtained from the soil bin.

Aw for T1 and T2 was calculated as a triangle shape by using Eq. (3.9) as shown in Fig. 3.15 (a).

$$Aw = \tfrac{1}{2}\,Wf \times d \qquad (3.9)$$

A_w for T3 and T4 was calculated as a triangle shape if the tine works at a depth less than d_1, and A_w is equal to the triangle plus trapezoidal shape if the tine works at a depth deeper than d_1 using Eq. (3.10) and (3.11) as shown in Fig. 3.15 (b) and (c).

$$Aw = Aw_1 + Aw_2 \qquad (3.10)$$
$$Aw = \tfrac{1}{2}\,Ww \times d_1 + \tfrac{1}{2}\,(Ww + Wf)d_2 \qquad (3.11)$$

Where: $Ww = Wt$ is the width of wing or width of tine in cm, respectively, d_1 is the wing height in cm, d_2 is the vertical distance from the wing to the soil surface and $d = d_1 + d_2$.

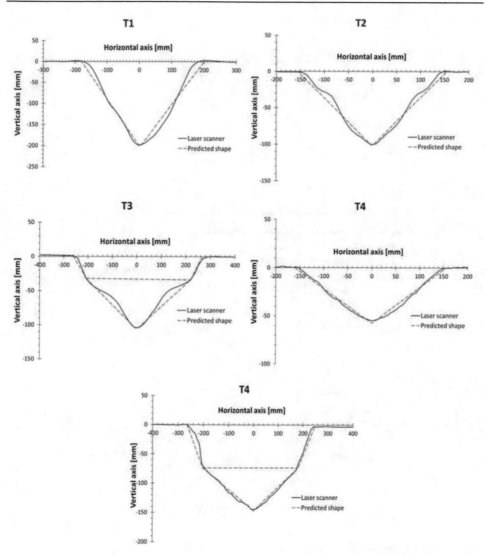

Fig. 3.14 Matching furrow shape

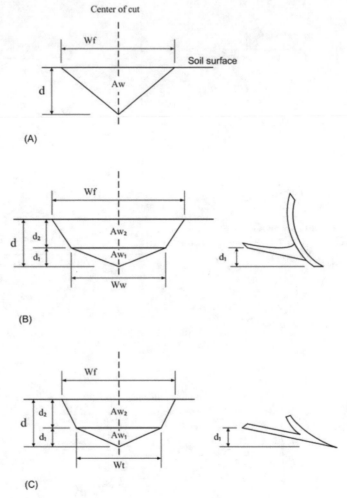

Fig. 3.15 Geometric furrow shapes

3.5.3 Soil Loosening Percentage

The soil loosening percentage $SLa\%$ was measured by using Eq. (3.12) (Mckyes & Maswaure 1997).

$$SLa\% = \frac{At - Aw}{At} \times 100 \qquad (3.12)$$

Where: At is the total area of disturbed soil. Eq. (3.12) can be written as

$$SLa\% = \frac{Ar}{At} \times 100 \qquad (3.13)$$

Where: Eq. (3.13) represents soil distribution percentage above the soil surface, without considering the soil distribution percentage under the soil surface ($SLu\%$), which can be represented by using Eq. (3.14).

$$SLu\% = \frac{Af}{At} \times 100 \qquad (3.14)$$

3.5.4 Specific Force and Specific Power

The specific force Fs per unit area in kN/m² was calculated by using Eq. (3.15).

$$Fs = \frac{Fd}{Aw} \tag{3.15}$$

Where Fd is the total force or draft force in kN and is presented by Eq. (3.16).

$$Fd = \sqrt{(Fh)^2 + (Fv)^2} \tag{3.16}$$

The specific power Ps per unit area in kW/m² was calculated by using Eq. (3.17).

$$Ps = Fs \times S \tag{3.17}$$

3.6 Field Test

To validate the result obtained from the soil bin for the draft force and their components and some of the soil profile properties, a field test was carried out using these four tines. A randomized complete block design was used (RCBD) at three speed levels 4, 12, and 20 km/h, three depths 5, 10, and 20 cm, and with three replicates, as shown in Fig. 3.16.

Fig. 3.16 Running the test in the field

The field has the same soil type as the bin, sandy loam. In addition, it was also weedy with stones and roots of the previous crop (see Fig. 3.17).

Fig. 3.17 Field test (51°22'40.1"N 13°26'07.3"E)

Soil moisture content and soil bulk density were measured during the test at 10.4% ± 1.1 (d.b.) and 1.42 g/cm³ ± 0.1, respectively. Manual measurements were done for d and Wf, as shown in Fig. 3.18.

Fig. 3.18 Measurements in the field for d and Wf

4 Results and Discussion

4.1 Force

4.1.1 Effects of Speed and Depth on the Horizontal Force

Figure 4.1 illustrates the effects of speed and depth on Fh for all tines in the soil bin conditions. From this figure, it can see that Fh increased with increasing speeds and depths for all tines.

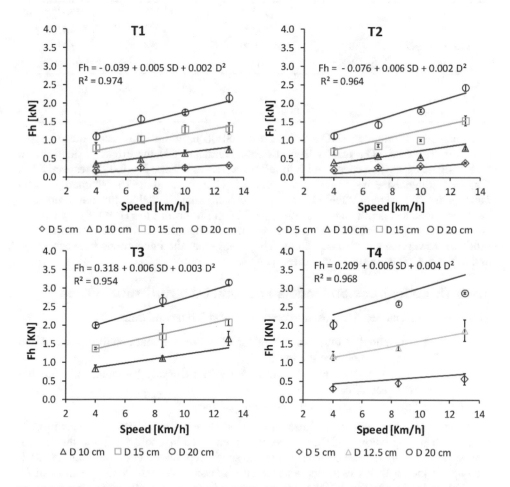

Fig. 4.1 Effects of speed and depth on Fh for all tines under soil bin conditions (mean ± SD)

The maximum values of Fh were 3.16, 2.88, 2.42 and 2.15 kN for T3, T4, T2 and T1, respectively, corresponding to a speed of 13 km/h at depth 20 cm, with absolute variations of 8.9%, 23.4%, and 31.9% between T3 and T4, T2 and T1, respectively. These differences are mainly due to T3 having a wider width than the other tines, which causes more soil to be cut and distributed.

© Springer-Verlag GmbH Germany, part of Springer Nature 2019
A. K. A. Al-Neama, *Evaluation of performance of selected tillage tines regarding quality of work*, Fortschritte Naturstofftechnik, https://doi.org/10.1007/978-3-662-57744-8_4

It can also be seen from this figure that the regression models for all tines referring to Fh increased linearly with the interaction term between speed - depth and quadratically with depth with high coefficients of determination $R^2 = 0.974$, 0.964, 0.954 and 0.968 for T1, T2, T3 and T4, respectively. The difference between the regression equations was in the coefficient of depth square, except T1 and T2 had only small differences in the coefficient of term speed-depth, namely 0.005 and 0.006, respectively.

Table 4.1 ANOVA table for the horizontal force

Source of variation	F – Value (Fh)			
	T1	T2	T3	T4
Speed	77.1**	276.7**	86.1**	48.3**
Depth	566.2***	1228.3***	218.6***	562.4***
Speed × Depth	9.8*	39.2*	2.5*	5.2*
R^2	0.984	0.993	0.959	0.986

* P < 0.01
** P < 0.001
*** P < 0.000

The analysis of variance (ANOVA table using SPSS program GLM procedure) is presented in Table 4.1. The results from Table 4.1 indicate that the effect of depth has a higher significance than speed on Fh for all tines ($P < 0.000$ and $P < 0.001$, respectively). Similar findings have been previously reported (Glancey et al. 1996, Rahman & Chen 2001, Sahu & Raheman 2006). It can also be seen from this table that the interaction between speed and depth had a significant effect on Fh for all tines ($P < 0.01$). Onwualu & Watts (1998) achieved different results when testing wide and narrow tines in silty sand soil at varying speeds and depths. They found that the interactions between speed and depth did not have significant effects on the draft force.

4.1.2 Three Regression Models to Predict the Horizontal Force

Three regression models were used to predict Fh at soil bin conditions.

- The first model is based on the effect of the operational conditions speed and depth for each tine (see section 3.5.1).
- The second model is based on using a statistic Dummy Variable. This variable represents each tine.
- The third model bases on the geometry of tines.

A stepwise selection with multi-linear regression at significance level 5% was used to evaluate these regression models. Results are summarized in Table 4.2. From this table it can be seen that Fh increased linearly with the speed-depth interaction term and quadratically with the depth for each tine and for all regression models (positive values of C_3 and C_5). It is obvious that all coefficients except C_5 increase with increasing width of tines in the operating condition model at a high coefficient of determinations $R^2 > 90\%$.

It can also be seen that the coefficient of interaction between speed-depth and the coefficient of depth squares C_3 and C_5 have the same value, namely 0.006 and 0.002, respectively, in the dummy and geometric model and are similar to the values in operating conditions. The similarity can be attributed to the stable test environment done in specific soil type and conditions.

Table 4.2 Regression models and coefficients for the horizontal force

Regression model	Tines		Regression coefficient							R^2
			C_0	C_1	C_2	C_3	C_4	C_5		
Operating condition	T1		- 0.039	n.s.	n.s.	0.005	n.s.	0.002		0.974
	T2		- 0.076	n.s.	n.s.	0.006	n.s.	0.002		0.964
	T3		0.318	n.s.	n.s.	0.006	n.s.	0.003		0.954
	T4		0.209	n.s.	n.s.	0.006	n.s.	0.004		0.968
		K	C_0	C_1	C_2	C_3	C_4	C_5		R^2
Dummy variable	T1	0.000								
	T2	0.008	- 0.166	n.s.	n.s.	0.006	n.s.	0.002		0.957
	T3	0.707								
	T4	0.574								
Geometric variable		G_1	G_2	G_3	C_0	C_1	C_2	C_3	C_4	C_5	R^2
	T	0.019	n.s.	n.s.	- 0.350	n.s.	n.s.	0.006	n.s.	0.002	0.952

n.s.: not significant

Table 4.2 shows that the coefficient K for a dummy regression model is equal to zero for T1 because it was set as the reference tine, and it can also be seen that the coefficient K increased with increasing width of tines at a high R^2 of 0.957.

As expected from previous regression models, only the coefficient of width G_1 appeared in the geometric regression model with a high value of 0.019 at a high R^2 of 0.952.

4.1.3 Validation of the Horizontal Force in Field Conditions

In order to verify the validity of the regression models obtained from the soil bin, the observed (field) and the predicted (regression) values of Fh for all tines are plotted in Fig. 4.2. It can be seen that the field test recorded higher values of Fh than predicted for all tines, which is due to the field conditions being different from the soil bin in regards to existence of stones, roots of the previous crop, and weeds that caused higher soil resistance.

From the comparison in Fig. 4.2, it can be seen that there is a general accordance of observed and predicted values of Fh for T1 with slope 0.744, 0.720 and 0.701 and with higher R^2 of 0.959, 0.960 and 0.954 for the dummy, geometric and operating condition regression, respectively. The variation between observed and predicted values of Fh was found to be 24% for the dummy, 25% for the geometric, and 29% for the operating condition regression.

A good approval between observed and predicted values was found for Fh for T2 plotted in Fig. 4.2 with slope 0.852, 0.844, and 0.816 and with higher R^2 of 0.966, 0.967 and 0.969 for the operating condition, geometric, and dummy regression, respectively. The variation between observed and predicted values of Fh was found to be 13% for both the operating condition and geometric and 19% for the dummy regression.

A general agreement between observed and predicted values was found for Fh for T3 (see Fig. 4.2) with slope 0.739, 0.719 and 0.709 and with higher R^2 of 0.933, 0.805 and 0.825 for the operating condition, dummy and geometric regression, respectively. The variation between observed and predicted values of Fh was found to be 18% for the operating condition, 16% for the dummy and 17% for the geometric regression.

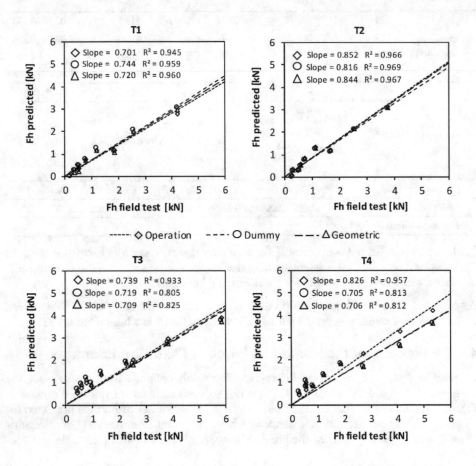

Fig. 4.2 Comparison between observed (field) and predicted (regression) for Fh for all tines

Tine T4 shows a general accordance of observed and predicted values of Fh in Fig. 4.2. From this graph, it can be seen that the slope of T4 is 0.826 for operation condition, 0.706 for geometric, and 0.705 for the dummy regression with a higher R^2 of 0.957, 0.813 and 0.812, respectively. The regression equation model predicted the Fh of the operating condition, geometric, and dummy regression with a variation 9%, 17% and 17%, respectively.

Note: The differences between predicted and observed for T2, T3 and T4 were less than 20%. The reason for this is that these tines are designed to operate in a weedy field, while T1 is designed to operate in heavy-duty conditions (dry and hard soil).

4.1.4 Effects of Speed and Depth on the Vertical Force

Effects of speed and depth on Fv for all tines were plotted in Fig. 4.3 under soil bin conditions. Note the sign of Fv refers to the direction, with (-) negative values forcing the tine to penetrate the soil and vice versa.

T1 and T2 had a positive value of Fv at a depth of 5 cm (see Fig. 4.3), and the direction of Fv changes with increasing depth, which is attributed to two factors. The first factor is related to the thickness of the tines and the second factor is tied to the curved shape of the tines. Therefore, it is not recommended to use these tines at shallow depths.

From this figure, it can be see that Fv increased with increasing speeds and depths for all tines. The maximum absolute values of Fv were 0.94, 0.84, 0.68 and 0.45 kN for T3, T4, T2 and T1, respectively, at speed 13 km/h and depth 20 cm, with absolute variations of 10.6%, 27.7% and 52.1% between T3 and T4, T2 and T1, respectively. These differences are mainly due to T3 being wider than other tines, which increased the contact area between tine and soil.

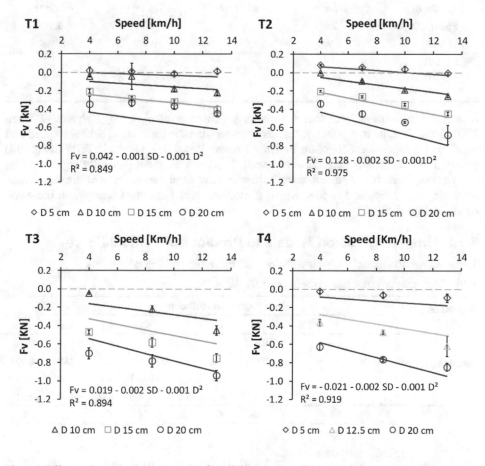

Fig. 4.3 Effects of speed and depth on Fv for all tines under soil bin conditions (mean ± SD)

It can also be seen from this figure that the regression models for all tines referring to Fv increased linearly with interaction term between speed-depth and quadratically with depth with high coefficients of determination R^2 of 0.849, 0.975, 0.894 and 0.919 for T1, T2, T3 and T4, respectively. A difference exists between T1 and other tines only for the coefficients of the interaction between speed-depth, where T1 contributes half the penetration force compared to T3. However, the coefficient of depth square has the same value 0.001 for all tines; this similarity can be attributed to the stable test environment done in specific soil type and conditions.

The analysis of variance (ANOVA table using SPSS program GLM procedure) is summarized in Table 4.3. From this table it can be seen that the effect of depth has a higher significance than speed on Fv for all tines ($P < 0.000$ and $P < 0.001$, respectively).

Table 4.3 ANOVA table for the vertical force

Source of variation	F – Value (Fv)			
	T1	T2	T3	T4
Speed	15.4**	120.7**	94.2**	31.5**
Depth	152.5***	671.1***	326.3***	467.3***
Speed × Depth	2.6*	6.4*	2.1*	3.5*
R²	0.943	0.987	0.970	0.983

* $P < 0.01$
** $P < 0.001$
*** $P < 0.000$

Identical results have been found previously (Glancey et al. 1996, Rahman & Chen 2001, Sahu & Raheman 2006). It can also be seen that the interaction between speed and depth had a significant effect on Fv for all tines ($P < 0.01$). Onwualu & Watts (1998) reported that the interactions between speed and depth were not generally significant on the vertical force for wide and narrow tines in silty sand soil under soil bin conditions with varying speeds and depths, which demonstrates the effect of changes in the environmental conditions.

4.1.5 Three Regression Models to Predict the Vertical Force

Three regression models were used to predict Fv in the soil bin conditions.

Table 4.4 Regression models and coefficients for the vertical force

Regression model	Tines				Regression coefficient						R^2
					C_0	C_1	C_2	C_3	C_4	C_5	
Operating condition	T1				0.042	n.s.	n.s.	- 0.001	n.s.	- 0.001	0.849
	T2				0.128	n.s.	n.s.	- 0.002	n.s.	- 0.001	0.957
	T3				0.019	n.s.	n.s.	- 0.002	n.s.	- 0.001	0.894
	T4				- 0.021	n.s.	n.s.	- 0.002	n.s.	- 0.001	0.919
		K			C_0	C_1	C_2	C_3	C_4	C_5	R^2
Dummy variable	T1	0.000									
	T2	- 0.030			0.144	n.s.	n.s.	- 0.002	n.s.	- 0.001	0.904
	T3	- 0.269									
	T4	- 0.226									
Geometric variable		G_1	G_2	G_3	C_0	C_1	C_2	C_3	C_4	C_5	R^2
	T	- 0.007	n.s.	n.s.	0.200	n.s.	n.s.	- 0.002	n.s.	- 0.001	0.903

n.s.: not significant

The first model is based on the effect of the operational conditions speed and depth for each tine (see section 3.5.1); the second model is based on using a statistic Dummy Variable. This variable represents each tine. The third model bases on the geometry of the tines.

A stepwise selection with multi-linear regression at significance level 5% was used to evaluate these regression models. Results are summarized in Table 4.4. From this table it can be seen that C_3 and C_5 had significant effects on Fv, the coefficient of interaction between speed-depth term and depth square, respectively, for all tines and all regression models.

It can also be seen that the coefficient of interaction between speed-depth and the coefficient of depth square C_3 and C_5 have the same absolute value of 0.002 and 0.001, respectively, in the dummy and geometric model and were the same as the values in operating conditions (except T1). The similarity can be attributed to the stable test environment done in specific soil type and conditions.

As anticipated from prior regression models, only the coefficient of width G_1 appeared in the geometric regression model with a high absolute value 0.007 at high R^2 of 0.903.

4.1.6 Validation of the Vertical Force in Field Conditions

With a view to verify the validity of the regression models obtained from the soil bin, the observed (field) values of Fv were plotted against their respective predicted (regression) values in Fig. 4.4 for all tines.

From the comparison in Fig. 4.4 it can be seen that there is a very good agreement between observed and predicted values of Fv for T1 with slope 0.906, 1.157 and 1.141 with higher R^2 of 0.925, 0.951 and 0.953 for the operating condition, dummy and geometric regression, respectively. The average absolute variation between observed and predicted values of Fv was found to be 7.4% for the operating condition, 16.7% for the geometric and 19.4% for the dummy regression.

An excellent correlation of the observed and the predicted values was found for Fv for T2 plotted in Fig. 4.4 with slope 0.986, 1.003, and 1.014 and with higher R^2 of 0.958, 0.988, and 0.941 for the operating condition, dummy and geometric regression, respectively. The average absolute variation between observed and predicted values of Fv was found to be 12.1% for the operating condition, 14.7% for the dummy and 17.1% for the geometric regression.

A general agreement of observed and predicted values exists for Fv for T3 in Fig. 4.4 with slope 0.666, 0.765 and 0.758 for the operating condition, dummy and geometric regression, respectively. The average absolute variation between observed and predicted values of Fv was found to be 16.6% for the operating condition, 22.4% for the dummy and 21.1% for the geometric regression.

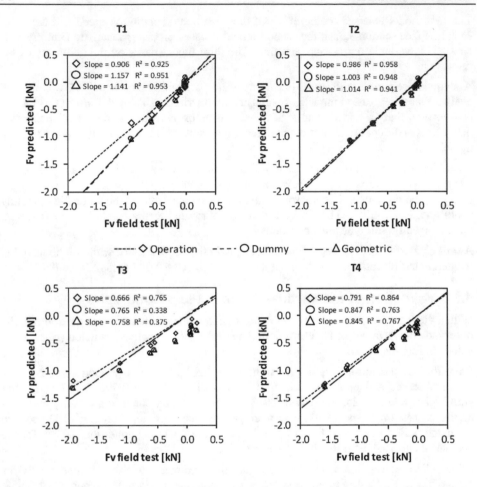

Fig. 4.4 Comparison between observed (field) and predicted (regression) for Fv for all tines

Tine T4 shows a general agreement of observed and predicted values of Fv in Fig. 4.4. From this chart, it can be seen that the slope of the trend line is 0.791 for operation condition, 0.847 for dummy and 0.845 for the geometric regression.

4.1.7 Effects of Speed and Depth on the Draft Force

The effects of speed and depth on Fd for all tines are presented in Fig. 4.5 under soil bin conditions. From this figure, it can be seen that Fd increased with increasing speeds and depths for all tines. The maximum values of Fd were 3.30, 3.00, 2.52 and 2.20 kN for T3, T4, T2 and T1, respectively, at speed 13 km/h and a depth of 20 cm, with absolute variations of 9.1%, 23.6% and 33.3% between T3 and other tines, respectively. These differences are mainly due to the T3 having a wider width than the other tines, which causes more soil to be cut and distributed as mentioned above for the horizontal force.

It can also be seen from this figure that the regression models for all tines referring to Fd increased linearly with interaction term between speed-depth and quadratically with depth with high coefficients of determination R^2 of 0.975, 0.976, 0.959 and 0.965 for T1, T2, T3 and T4, respectively. The differences between the regression equations were in the coefficient of interaction between speed-depth (except T2 and T4) and in the coefficient of depth square (except T1 and T2).

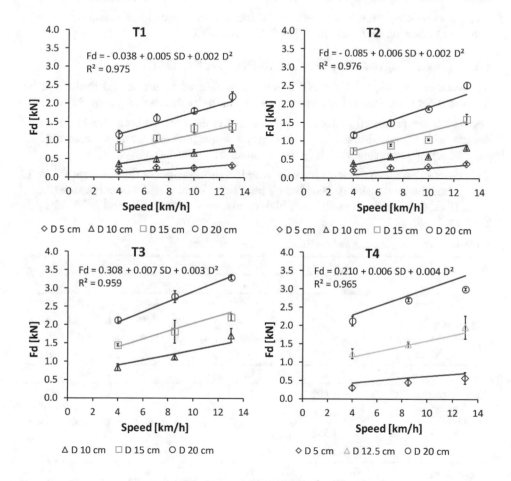

Fig. 4.5 Effects of speed and depth on Fd for all tines under soil bin conditions (mean ± SD)

Table 4.5 ANOVA table for the draft force

Source of variation	F – Value (Fd)			
	T1	T2	T3	T4
Speed	78.9**	300.1**	95.2**	46.4**
Depth	588.4***	1364.2***	248.1***	550.8***
Speed × Depth	9.6*	42.7*	2.1	4.9*
R^2	0.985	0.994	0.975	0.985

* P < 0.01
** P < 0.001
*** P < 0.000

The analysis of variance (ANOVA table using SPSS program GLM procedure) is presented in Table 4.5.The results of Table 4.5 indicated that the effect of depth has a higher significance than speed on Fd for all tines ($P < 0.000$ and $P < 0.001$, respectively).

It can also be seen from this table that the interaction between speed and depth had a significant effect on Fd for all tines ($P < 0.01$) except T3.

4.1.8 Three Regression Models to Predict the Draft Force

As mentioned above, three regression models were used to predict Fd in the soil bin conditions for the horizontal and vertical forces (for more details see section 3.5.1).

The results are presented in Table 4.6. From this table it can be seen that Fd increased linearly with the speed-depth interaction term and quadratically with the depth for each tine and for all regression models (positive values of C_3 and C_5).

It is obvious that all coefficients increase with increasing width of tines, except C_3 for T2 and T3 had the same value (0.006) and C_5 for T1 and T2 also had the same value (0.002) in the operating condition model with high coefficients of determination ($R^2 > 90\%$) for all tines.

Table 4.6 Regression models and coefficients for the draft force

Regression model	Tines			Regression coefficient						R^2
				C_0	C_1	C_2	C_3	C_4	C_5	
Operating condition	T1			- 0.038	n.s.	n.s.	0.005	n.s.	0.002	0.975
	T2			- 0.085	n.s.	n.s.	0.006	n.s.	0.002	0.976
	T3			0.308	n.s.	n.s.	0.007	n.s.	0.003	0.959
	T4			0.210	n.s.	n.s.	0.006	n.s.	0.004	0.965
			K	C_0	C_1	C_2	C_3	C_4	C_5	R^2
Dummy variable	T1		0.000							
	T2		0.018	- 0.185	n.s.	n.s.	0.006	n.s.	0.003	0.956
	T3		0.752							
	T4		0.607							
Geometric variable		G_1	G_2	G_3	C_0	C_1	C_2	C_3	C_4	C_5	R^2
	T	0.020	n.s.	n.s.	- 0.375	n.s.	n.s.	0.006	n.s.	0.003	0.952

n.s.: not significant

It can also be seen from this table that the coefficient of interaction between speed-depth and the coefficient of depth square C_3 and C_5 have the same value (0.006 and 0.003, respectively) in the dummy and geometric models and are similar to the values in operating conditions.

The similarity can be attributed to the stable test environment done in specific soil type and conditions. Table 4.6 shows that the coefficient K for a dummy regression model is equal to zero for T1 because it was set as tine reference, and it can also be seen that the coefficient K increased with increasing width of tines at high R^2 of 0.956.

Only the coefficient of width G_1 appeared in the geometric regression model with a high value of 0.020 at high R^2 of 0.952 (see Table 4.6).

4.1.9 Validation of the Draft Force in Field Conditions

To prove the exactitude of the regression models obtained from the soil bin, the observed (field) and the predicted (regression) values of Fd are plotted in Fig. 4.6 for all tines.

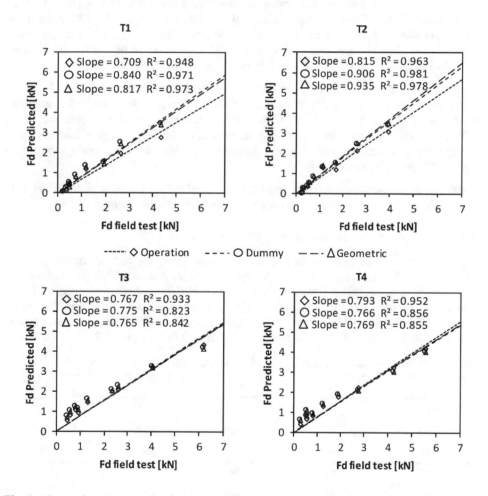

Fig. 4.6 Comparison between the observed (field) and predicted (regression) for Fd for all tines

From this figure, it can be seen that the field test recorded higher values of Fd than predicted for all tines, which is due to the differences between the field and the soil bin (as mentioned above for the horizontal force).

From the comparison in Fig. 4.6 it turned out that there is a good accordance of observed and predicted values of Fd for T1 with slope 0.840, 0.817 and 0.709 and with higher R^2 of 0.971, 0.973 and 0.948 for the dummy, geometric and operating condition regression, respectively. The average variation between observed and predicted values of Fd was found to be 14% for the dummy, 19% for the geometric and 23% for the operating condition regression.

An excellent acceptance between observed and predicted values for Fd was found (Fig. 4.6) for T2 with slope 0.935, 0.906, and 0.815 at higher R^2 of 0.978, 0.981 and 0.963 for the geometric, dummy and operating condition regression, respectively. With average variation between observed and predicted values were 4% for the geometric, 8% for the dummy and 15% for the operating condition regression.

Figure 4.6 shows a good agreement between observed and predicted values for Fd for T3 with slope 0.775, 0.767 and 0.765 at higher R² of 0.823, 0.933, and 0.842 for the dummy, operating condition and the geometric regression, respectively. The average variation between observed and predicted values of Fd were found to be 9% for the dummy, 11% for the geometric and 15% for the operating condition regression.

Tine T4 shows a good assent between observed and predicted values of Fd (see Fig. 4.6). From this graph, it can see that the slopes of the best-fit line were 0.793 for operation condition, 0.769 for geometric, and 0.766 for the dummy regression with higher R² of 0.952, 0.855 and 0.856, respectively. The regression equation model predicted the Fd of the operating condition, geometric and dummy regression with the average variation 13%, 11% and 11%, respectively.

Note: The differences between predicted and observed values for Fd were less than 20% for all tines and all regression models (except T1 with operation condition 23%).

4.1.10 Correlation between Fd, Fh and Fv

The matrix correlation between draft force, horizontal force and vertical force is presented in Table 4.7. From this table it can be seen that there is a very high positive significant correlation coefficient between Fd and Fh ($P < 0.01$), which amounted to 1.000, 1.000, 0.999, and 1.000 for T1, T2, T3, and T4, respectively, a negative correlation between Fd and Fv of -0.918, -0.965, -0.946 and -0.991, respectively, and negative correlation between Fh and Fv of -0.991, -0.961, -0.934 and -0.989, respectively.

Table 4.7 Matrix correlation between draft force, horizontal force and vertical force

Parameter	Tine	Fd	Fh	Fv
Fd	T1	1.000		
Fh	T1	1.000*	1.000	
Fv	T1	- 0.918*	- 0.911*	1.000
Fd	T2	1.000		
Fh	T2	1.000*	1.000	
Fv	T2	- 0.965*	- 0.961*	1.000
Fd	T3	1.000		
Fh	T3	0.999*	1.000	
Fv	T3	- 0.946*	- 0.934*	1.000
Fd	T4	1.000		
Fh	T4	1.000*	1.000	
Fv	T4	- 0.991*	- 0.989*	1.000

* Statistical significance at P-value < 0.01

The comparison between regression equation for Fd and Fh for all tines is presented in Table 4.8. From this table it can be seen that the coefficient of interaction between speed-depth had the same values in the Fd and Fh regression equations for all tines (except T3 had a small difference between Fd and Fh of 0.007 and 0.006, respectively).

In addition, the coefficient of depth square had the same values in the regression equations for Fd and Fh for all tines. Similar values were obtained from the regression equations for Fd and Fh for all tines for the intercept coefficient.

Table 4.8 Comparison between regression equations for Fd and Fh for all tines

Tines	Regression model			
	Fd	R^2	Fh	R^2
T1	- 0.038 + 0.005 SD + 0.002 D^2	0.975	- 0.039 + 0.005 SD + 0.002 D^2	0.974
T2	- 0.085 + 0.006 SD + 0.002 D^2	0.967	- 0.076 + 0.006 SD + 0.002 D^2	0.964
T3	0.308 + 0.007 SD + 0.003 D^2	0.959	0.318 + 0.006 SD + 0.003 D^2	0.954
T4	0.210 + 0.006 SD + 0.004 D^2	0.965	0.209 + 0.006 SD + 0.004 D^2	0.968

Table 4.9 shows the regression equation to predict Fd from Fh or vice versa. From this table it can be seen that the coefficient of Fh to predict Fd is equal to the number one with three decimal points, and it is approximately equal to one integer for all tines. Besides, the coefficient of intercept is very small with a negative value (except T1) and can be neglected for all tines.

It can also be seen from Table 4.9 that the coefficient of Fd to predict Fh is smaller than number one with three decimal points, it is approximately equal to one integer for all tines, and the coefficient of intercept is very small with a positive value (except T1) and can be neglected for all tines.

Table 4.9 Regression equation to predict Fd from Fh or vice versa

tines	Regression equation for (Fd)	R^2	tines	Regression equation for (Fh)	R^2
T1	0.003 + 1.028 Fh	1.000	T1	- 0.002 + 0.973 Fd	1.000
T2	- 0.004 + 1.046 Fh	1.000	T2	0.004 + 0.956 Fd	1.000
T3	- 0.021 + 1.057 Fh	0.999	T3	0.022 + 0.945 Fd	0.999
T4	- 0.011 + 1.049 Fh	1.000	T4	0.011 + 0.953 Fd	1.000

Therefore, as mentioned above, it can be said that, statistically, Fd is equal to Fh for all the tines used in the sandy loam soil.

4.2 Soil Profile

4.2.1 Distance

4.2.1.1 Effect of Speed and Depth on the Height of the Furrow

Effects of speed and depth on the HF in cm for all tines were plotted in Fig. 4.7 under soil bin conditions. From this figure and the regression model, it can be seen that HF characteristic was affected differently by speed and depth for each tine. The HF for T1 increased with speed only. If the depth of 5 cm was neglected as recommended (see section 4.1.4), there is no effect of depth on HF. Meanwhile, the HF for T2 increased with speed and depth (Fig. 4.7). It can be seen from Fig. 4.7 that the HF for T3 is not affected by speed or depth. However, HF increases with depth only for T4 (Fig. 4.7) until a depth of 12.5 cm and then it decreases. The maximum values of HF were 11.7, 8.5, 5.6 and 2.6 cm for T2, T1, T4 and T3, respectively. Variations of 27%, 52% and 78% were found between T2 and other tines, respectively.

Note: Based on the results, T2 can be used as a furrow opener in seeding equipment.

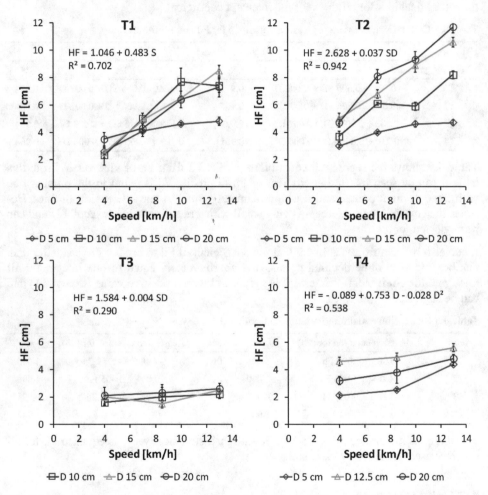

Fig. 4.7 Effects of speed and depth on HF for all tines under soil bin conditions (mean ± SD)

The analysis of variance (ANOVA table using SPSS program GLM procedure) is presented in Table 4.10. The results from Table 4.10 indicated that the effect of speed has a higher significance than depth on HF for T1 ($P < 0.01$ and $P < 0.05$, respectively).

Both speed and depth significantly affected HF for T2 and T4. While only depth affected HF for T3. In addition, it can be seen from this table that the interaction between speed and depth had a significant effect on HF for T1, T2, and T4 ($P < 0.05$), but not T3.

Table 4.10 ANOVA table for the height of the furrow

Source of variation	F – Value (HF)			
	T1	T2	T3	T4
Speed	178.8**	410.4***	1.8	9.6*
Depth	28.8*	415.2***	4.7*	23.0*
Speed × Depth	9.1*	26.4*	2.0	9.7*
R²	0.957	0.998	0.540	0.853

* P < 0.05
** P < 0.01
*** P < 0.001

4.2.1.2 Effect of Speed and Depth on the Height of the Ridge

Effects of speed and depth on HR in cm are presented in Fig. 4.8 for all tines under the soil bin conditions.

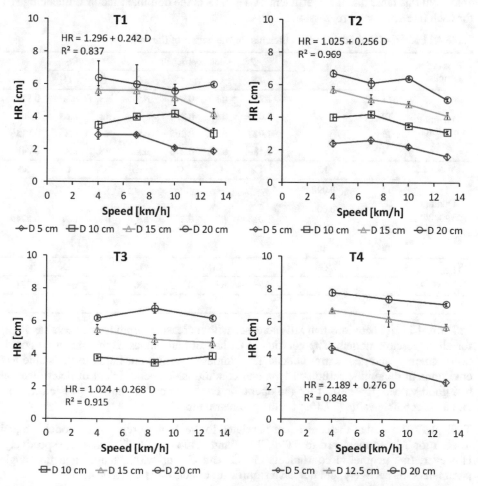

Fig. 4.8 Effects of speed and depth on HR for all tines under soil bin conditions (mean ± SD)

From this figure and the regression model, it can be seen that a similar manner is attained in this characteristic for every tine. The HR increased only with depth for all tines with maximum values of 7.8, 6.8, 6.7 and 6.4 cm for T4, T3, T2 and T1, respectively, with a variation of 13%, 14% and 18% between T4 and other tines, respectively. The regression equations gave a good fit with an R^2 of 0.969, 0.915, 0.848 and 0.837 for T2, T3, T4 and T1, respectively; the high values of R^2 indicate that the variable depth in the regression equation can explain most of the variability in the test data. It is obvious that the coefficient of depth in the regression equation has almost the same value for each tine.

Note: Based on the results, there is no effect when adding a wing (increasing tine width) to T2 for this parameter.

4.2.1.3 Three Regression Models to Predict the Height of the Ridge

Three regression models were used to predict HR in the soil bin conditions (as mentioned above) for the forces section. The results are presented in Table 4.11. It is noticeable from this table that the coefficient of depth C_2 is the dominant factor influencing HR for each tine and for all regression models.

Table 4.11 Regression models and coefficients for the height of the ridge

Regression model	Tines				Regression coefficient						R^2
					C_0	C_1	C_2	C_3	C_4	C_5	
Operating condition	T1				1.296	n.s.	0.242	n.s.	n.s.	n.s.	0.837
	T2				1.025	n.s.	0.256	n.s.	n.s.	n.s.	0.969
	T3				1.024	n.s.	0.268	n.s.	n.s.	n.s.	0.915
	T4				2.189	n.s.	0.276	n.s.	n.s.	n.s.	0.848
			K		C_0	C_1	C_2	C_3	C_4	C_5	R^2
Dummy variable	T1		0.090								
	T2		0.000		1.016	n.s.	0.257	n.s.	n.s.	n.s.	0.874
	T3		0.172								
	T4		1.406								
Geometric variable		G_1	G_2	G_3	C_0	C_1	C_2	C_3	C_4	C_5	R^2
	T	n.s.	n.s.	n.s.	1.356	n.s.	0.255	n.s.	n.s.	n.s.	0.771

n.s.: not significant

In Table 4.11, it is obvious that HR increases with increasing depth (positive value of C_2) for all regression models. The coefficient C_2 has similar values in the operating conditions, dummy, and geometric models. The similarity can be attributed to the stable test environment done in specific soil type and conditions. The coefficient of determination had good values ($R^2 > 80\%$) for the operating condition and dummy regression models and an acceptable value (77.1%) for the geometric model.

Table 4.11 shows that the coefficient K related to the dummy regression model is equal to zero for T2 and increased to 0.090, 0.172 and 1,406 for T1, T3 and T4, respectively. However, tine geometric coefficients G_1, G_2 and G_3 for width, length and tine angle parameters, respectively, did not have significant effects on HR (Table 4.11).

4.2.1.4 Validation of the Regression Models for HR in Soil Bin Conditions

To substantiate the reliability of the regression models for HR, the observed values (soil bin) were plotted in Fig. 4.9 against the predicted values (regression) for all tines. From this figure, an excellent accordance between soil bin and regression models for HR for all tines can be seen.

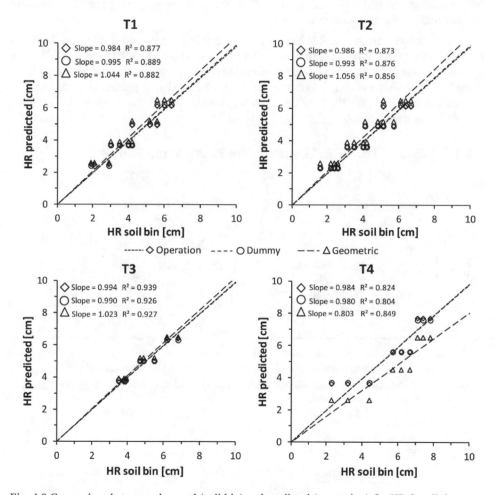

Fig. 4.9 Comparison between observed (soil bin) and predicted (regression) for HR for all tines

Tine T1 shows an excellent correlation between observed and predicted values of HR (Fig. 4.9). From this graph, it can be seen that the slopes of the best fit line were 1.044 for geometric, 0.995 for dummy and 0.984 for the operation condition at higher $R^2 >$ 80% for all regression models. The regression equation model predicted the HR of the dummy, operating condition and geometric regression with the average absolute variation of 1%, 2% and 7%, respectively.

Figure 4.9 shows an excellent agreement between observed and predicted values of HR for T2 with slope 1.056, 0.993 and 0.986 for the geometric, dummy and operating condition regressions, respectively, at higher $R^2 >$ 80%. The average absolute variation between observed and predicted values of HR were found to be 2% for the operating condition, 3% for the dummy and 11% for the geometric regression.

An excellent acceptance between observed and predicted values of HR was found (Fig. 4.9) for T3 with slope 1.023, 0.994 and 0.990 for the geometric, dummy and operating condition regressions, respectively, at higher $R^2 > 90\%$, with average absolute variation between observed and predicted values of 2% for both the operating condition and the dummy and 3% for the geometric regression.

From the comparison in Fig. 4.9 it is clear that there is a very good accordance between observed and predicted values of HR for T4 with slope 0.984, 0.980 and 0.803 for the operating condition, dummy and geometric regressions, respectively, and with higher $R^2 > 80\%$. The average absolute variation between observed and predicted values of HR was calculated to be 4% for both the operating condition and the dummy and 18% for the geometric regressions.

4.2.1.5 Effect of Speed and Depth on the Width of the Furrow

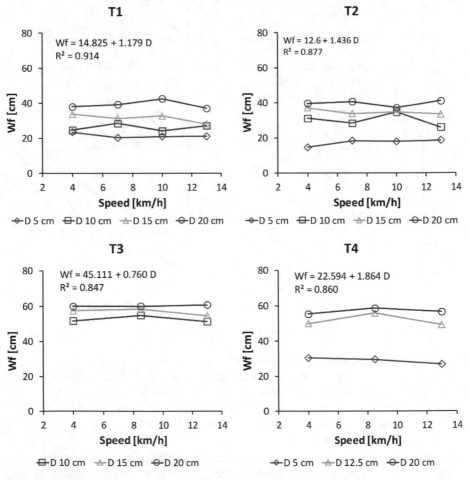

Fig. 4.10 Effects of speed and depth on Wf for all tines under soil bin conditions

Figure 4.10 illustrates the effect of speed and depth on the Wf in cm for all tines under soil bin conditions. From this figure and the regression model, it can be seen that a similar manner is attained in this characteristic for every tine.

The Wf increased only with depth for all tines with maximum value of 60.7, 58.7, 42.6 and 41.2 cm at the deeper operating depth of 20 cm for T3, T4, T1 and T2, respectively, with variations of 3%, 30% and 32% between T3 and other tines, respectively.

The regression equations provide a good fit with a coefficient of determination R^2 equal to 0.914, 0.8.77, 0.860 and 0.847 for T1, T2, T4 and T3, respectively; the high values of R^2 indicate that the variable depth in the regression equation can explain most of the variability in the experiment data.

4.2.1.6 Three Regression Models to Predict the Width of the Furrow

Three regression models were used to predict Wf in the soil bin conditions (as mentioned above) for the forces section. The results are summarized in Table 4.12. It is obvious from this table that the coefficient of depth C_2 is the main factor influencing Wf for each tine and for all regression models.

Table 4.12 Regression models and coefficients for the width of the furrow

Regression model	Tines				Regression coefficient						R^2
					C_0	C_1	C_2	C_3	C_4	C_5	
Operating condition	T1				14.825	n.s.	1.179	n.s.	n.s.	n.s.	0.914
	T2				12.600	n.s.	1.436	n.s.	n.s.	n.s.	0.877
	T3				45.111	n.s.	0.760	n.s.	n.s.	n.s.	0.847
	T4				22.594	n.s.	1.864	n.s.	n.s.	n.s.	0.860
			K		C_0	C_1	C_2	C_3	C_4	C_5	R^2
Dummy variable	T1		0.000								
	T2		0.988		12.638	n.s.	1.354	n.s.	n.s.	n.s.	0.929
	T3		24.034								
	T4		16.331								
Geometric variable		G_1	G_2	G_3	C_0	C_1	C_2	C_3	C_4	C_5	R^2
	T	0.598	n.s.	n.s.	6.761	n.s.	1.391	n.s.	n.s.	n.s.	0.910

n.s.: not significant

In Table 4.12, it is clear that Wf increases with increasing depth (positive value of C_2) for all regression models, with good values for the coefficient of determination ($R^2 >$ 80%) for operating condition and $R^2 > 90\%$ for both dummy and geometric models.

Table 4.12 shows that the coefficient K for a dummy regression model is equal to zero for T1 because it was set as the reference tine and it increased to 0.988, 16.331, and 24.034 for T2, T4, and T3, respectively, with increasing tine width.

As expected from the previous regression, only the coefficient of tine width G_1 appeared in the geometric regression model with a high $R^2 > 90\%$ (Table 4.12).

4.2.1.7 Validation of the Regression Models for the Wf in the Field

In order to confirm the veracity of the regression models obtained from the soil bin, the observed (field, manual measuring) and the predicted (regression, laser scanner measuring) values of Wf for all tines are plotted in Fig. 4.11. It can be seen from this figure that the regression models of the predictions had higher values of Wf than the field tests for T1 and T2. Which is due to the field condition being different to the soil bin in regard to the existence of stones, roots of the previous crop and weeds that lead to an increase of the overall cohesion of the soil, which causes a reduction of soil rift rate.

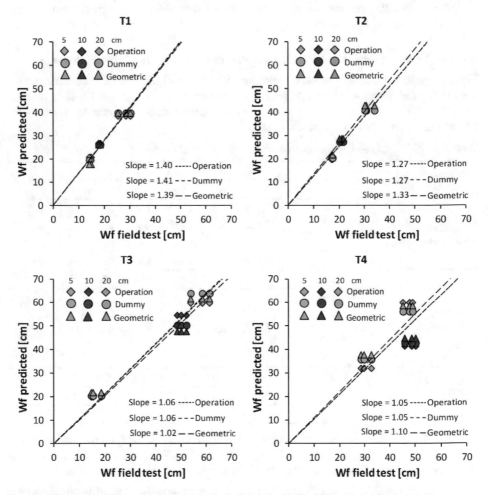

Fig. 4.11 Correlation between observed and predicted for Wf at different depths for all tines

Tine T1 shows a wide correlation between observed and predicted values of Wf (Fig. 4.11). From this graphic, it can see that the slopes of the best-fit line were 1.39 for geometric, 1.40 for operation condition and 1.41 for the dummy. The regression equation model predicted the Wf of the geometric, operating condition and dummy regression with average absolute variations 38%, 41% and 41%, respectively.

Figure 4.11 shows a wide range between observed and predicted values of Wf for T2 with slope 1.27 for both operating condition and dummy and 1.33 for the geometric regression. The average absolute variation between observed and predicted values of Wf were found to be 25% for the operating condition, 26% for the dummy and33 % for the geometric regression.

An excellent acceptance between observed and predicted values of Wf was found (Fig. 4.11) for T3 with slope 1.02 for the geometric and 1.06 for both the dummy and operating condition regressions, with average absolute variations between observed and predicted values of 5%, 7% and 8 % for the geometric, operating condition and the dummy regression, respectively.

From the comparison in Fig. 4.11, it is clear that there is a very good accordance between observed and predicted values of Wf for T4 with slope 1.05 for both the operating condition and the dummy and 1.10 for the geometric regressions. The average absolute variation between observed and predicted values of Wf was calculated to be 5%, 6% and 12% for operating condition, dummy and geometric regressions, respectively.

Note: With these results in mind, T3 and T4 are best for weed control and to cut the roots of the previous crop.

4.2.1.8 Effect of Speed and Depth on the Ridge to Ridge Distance

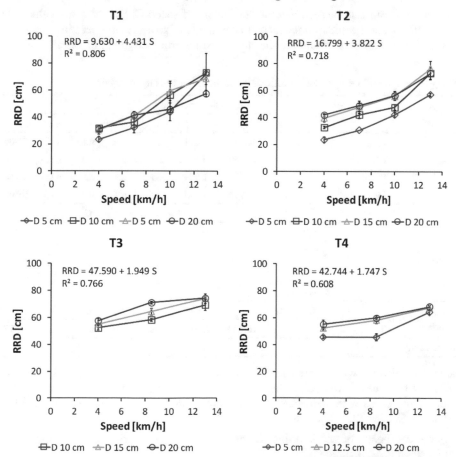

Fig. 4.12 Effects of speed and depth on RRD for all tines under soil bin conditions (mean ± SD)

Effects of speed and depth on the RRD in cm are plotted in Fig. 4.12 for all tines under soil bin conditions. From this figure and also the regression model it can be seen that a similar manner is attained in this characteristic for every tine.

The RRD increased with speed only for all tines with maximum values of 76.5, 74.4, 73.0 and 68.5 cm for T2, T3, T1 and T4, respectively, at a speed of 13 km/h, while variation was 3%, 5% and 10% between T2 and other tines, respectively.

The regression equations provided a good fit with R^2 of 0.806, 0.766, 0.718 and 0.608 for T1, T3, T2 and T4, respectively; the high values of R^2 indicate that the variable speed in the regression equation can explain most of the variability in the test data.

4.2.1.9 Three Regression Models to Predict the Ridge to Ridge Distance

Three regression models were used to predict RRD in the soil bin conditions (as mentioned above) for the forces section. The results are presented in Table 4.13. It is obvious from this table that the coefficient of speed C_1 is the predominant factor influencing RRD for each tine and for all regression models. From this table it can be seen that RRD increased linearly with the speed for each tine and for all regression models (positive values of C_1).

In addition, it can also be seen that the coefficient of speed C_1 has the same value 3.181 in the dummy and geometric models and is similar to the values in operating conditions. This similarity can be attributed to the stable test environment done in specific soil type and conditions.

Table 4.13 Regression models and coefficients for the ridge-to-ridge distance

Regression model	Tines				Regression coefficient						R^2
					C_0	C_1	C_2	C_3	C_4	C_5	
Operating condition	T1				9.630	4.431	n.s.	n.s.	n.s.	n.s.	0.806
	T2				16.799	3.822	n.s.	n.s.	n.s.	n.s.	0.718
	T3				47.590	1.949	n.s.	n.s.	n.s.	n.s.	0.766
	T4				42.744	1.747	n.s.	n.s.	n.s.	n.s.	0.608
			K		C_0	C_1	C_2	C_3	C_4	C_5	R^2
Dummy variable	T1		0.000								
	T2		2.763		19.483	3.181	n.s.	n.s.	n.s.	n.s.	0.736
	T3		17.634								
	T4		11.068								
Geometric variable		G_1	G_2	G_3	C_0	C_1	C_2	C_3	C_4	C_5	R^2
	T	0.407	n.s.	n.s.	16.700	3.181	n.s.	n.s.	n.s.	n.s.	0.727

n.s.: not significant

Table 4.13 also shows that the coefficient K for a dummy regression model is equal to zero for T1 because it was set as the reference tine, and it can also be seen that the coefficient K increased with increasing width of tines from 0.000 to 2.763 and to 11.068 and then to 17.634 for T1, T2, T4 and T3, respectively.

Only the coefficient of width of tine G_1 appeared in the geometric regression model with a good R^2 of 0.727 (Table 4.13).

4.2.1.10 Validation of the Regression Models for RRD in the Soil Bin

To verify the accuracy of the regression models for RRD, the observed values (soil bin) and the predicted values (regression models) are plotted in Fig. 4.13 for all tines. From this figure, an excellent correlation between observed and predicted values of RRD for all tines can be seen.

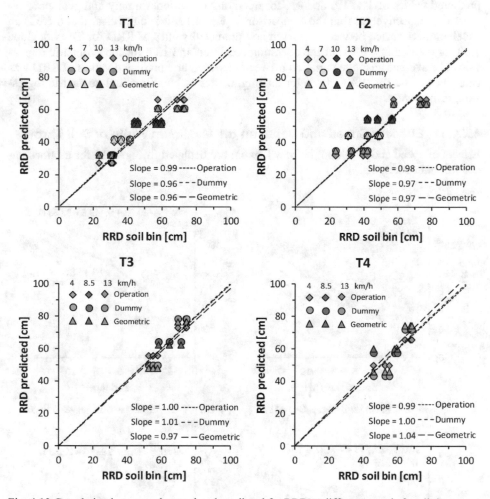

Fig. 4.13 Correlation between observed and predicted for RRD at different speeds for all tines

T1 shows an excellent accordance between observed and predicted values of RRD (Fig. 4.13). From this graphic, it can see that the slopes of the best-fit line were 0.99 for operation condition and 0.96 for both dummy and geometric regression. The regression equation model predicted the RRD of the operating condition, geometric and dummy regression with average absolute variations of 1%, 4% and 4%, respectively.

Figure 4.13 shows an excellent agreement between observed and predicted values of RRD for T2 with slope 0.98 for operating condition and 0.97 for both dummy and geometric regression. The average absolute variations between observed and predicted values of RRD were found to be 3% for the operating condition and 4% for the dummy and geometric regression.

An excellent acceptance between observed and predicted values of RRD was found (Fig. 4.13) for T3 with slope 1.00 for the operating condition, 1.01 for the dummy, and 0.97 for the geometric regression, with average absolute variation between observed and predicted values of 1%, 1% and 4% for operating condition, dummy and geometric regression, respectively. From the comparison in Fig. 4.13, it is quite clear that there is an excellent accordance between observed and predicted values of RRD for T4 with slope 1.00 for the dummy, 0.99 for the operating condition and 1.04 for the geometric regression. The average absolute variation between observed and predicted values of RRD was calculated to be 1%, 1% and 4% for dummy, operating condition and geometric regression, respectively.

4.2.1.11 Effect of Speed and Depth on the Maximum Width of Soil Throw

Effects of speed and depth on the MWT in cm are depicted in Fig. 4.14 for all tines under soil bin conditions.

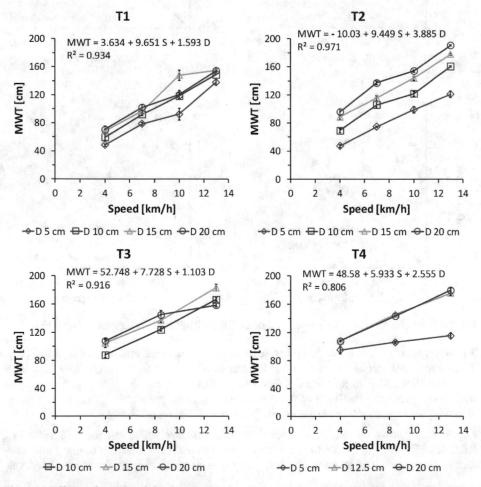

Fig. 4.14 Effects of speed and depth on MWT for all tines under soil bin conditions (mean ± SD)

From this figure and the regression model, it can be seen that a similar manner is attained in this characteristic for every tine.

The MWT increased with increasing speed and depth for all tines. The maximum values were 190.5, 183.2, 180 and 154.9 cm for T2, T3, T4 and T1, respectively, with variations of 4%, 6% and 19% between T2 and other tines, respectively. T1 is affected more than other tines by speed, while T2 is affected more than other tines by depth (higher value for the coefficient of speed and depth in the regression equation, respectively).

Note: According to the result, there is no effect when adding a wing to T2, while T2, T3 and T4 are useful for weed control since it covers the weeds with the soil.

The analysis of variance is presented in Table 4.14. It can be seen from this table that the effect of speed is of higher significance than depth on MWT for T1 ($P < 0.000$ and $P < 0.05$, respectively), T2 ($P < 0.000$ and $P < 0.001$, respectively), and T3 ($P < 0.001$ and $P < 0.01$, respectively).

Table 4.14 ANOVA table for the maximum width of the soil throw

Source of variation	F – Value (MWT)			
	T1	T2	T3	T4
Speed	928.9***	1450.2***	636.7**	781.3**
Depth	96.2*	743.1**	33.4*	528.0**
Speed × Depth	11.1*	6.5*	16.9*	74.8*
R^2	0.990	0.995	0.987	0.994

* $P < 0.05$ and 0.01
** $P < 0.001$
*** $P < 0.000$

The same significant effects for speed and depth were found on MWT for T4. It can also be seen from this table that the interaction between speed and depth had a significant effect on MWT for all tines ($P < 0.05$ and 0.01).

4.2.1.12 Three Regression Models to Predict the MWT

Three regression models were used to predict MWT in the soil bin conditions (as mentioned above) from the previous section. The results are presented in Table 4.15. It is clear from this table that MWT increased linearly with speed and depth for each tine and for all regression models (positive values of C_1 and C_2).

The coefficient of speed C_1 has an equivalent value of 8.454 in the dummy and geometric model; in addition, the coefficient of depth C_2 has the same value in the dummy and geometric model (2.532 and 2.522, respectively) and has similar values in operating conditions. This similarity can be attributed to the fixed experiment environment done in a specific soil type and conditions.

Table 4.15 shows that the coefficient K for a dummy regression model is equal to zero for T4 because it was set as the reference tine, and it can also be seen that the coefficient K decreased with increased tine length (see Table 3.2 in chapter 3, section 3.2).

Only the coefficient of length of tine G_2 appeared in the geometric regression model with a negative value of -1.305 at $R^2 > 80\%$ (Table 4.15).

Table 4.15 Regression models and coefficients for the maximum width of soil throw

Regression model	Tines			Regression coefficient						R^2
				C_0	C_1	C_2	C_3	C_4	C_5	
Operating condition	T1			3.634	9.651	1.593	n.s.	n.s.	n.s.	0.934
	T2			-10.030	9.449	3.885	n.s.	n.s.	n.s.	0.971
	T3			52.784	7.728	1.103	n.s.	n.s.	n.s.	0.916
	T4			48.580	5.933	2.555	n.s.	n.s.	n.s.	0.806
			K	C_0	C_1	C_2	C_3	C_4	C_5	R^2
Dummy variable	T1		- 25.367							
	T2		- 12.100	27.434	8.454	2.532	n.s.	n.s.	n.s.	0.892
	T3		- 2.290							
	T4		0.000							
Geometric variable		G_1	G_2	G_3	C_0	C_1	C_2	C_3	C_4	C_5	R^2
	T	n.s.	-1.305	n.s.	67.740	8.454	2.522	n.s.	n.s.	n.s.	0.881

n.s.: not significant

4.2.1.13 Validation of the Regression Models for the MWT in the Soil Bin

To verify the accuracy of the regression models for MWT, the observed values (soil bin) are plotted against the predicted values (regression models) in Fig. 4.15 for all tines. From this comparison, an excellent correlation between observed and predicted values of MWT for all tines can be seen.

T1 shows an excellent agreement between observed and predicted values of MWT (Fig. 4.15). From this figure, it can be seen that the slopes of the best-fit line were 0.995, 0.988 and 1.024 for operation condition, dummy, and geometric regression, respectively. The regression equation model predicted the MWT of the operating condition, dummy and geometric regression with average absolute variations of 1%, 1% and 6%, respectively.

Figure 4.15 shows an excellent correlation between observed and predicted values of MWT for T2 with slope 0.999 for operating condition, 0.980 for dummy, and 0.941 for geometric regression. The average absolute variations between observed and predicted values of MWT were found to be 1% for the operating condition, 3% for the dummy and 2% for the geometric regression.

An excellent acceptance between observed and predicted values of MWT was found (Fig. 4.15) for T3 with slope 0.997 for the operating condition, 1.002 for the dummy and 1.007 for the geometric regression, with the average absolute variation between observed and predicted values of 1% for all regression models.

From the comparison in Fig. 4.15, it is clear that there is an excellent accordance between observed and predicted values of MWT for T4 with slope 0.991 for the operating condition, 1.002 for the dummy and 1.009 for the geometric regression. The average absolute variation between observed and predicted values of MWT was calculated to be 1% for all regression models.

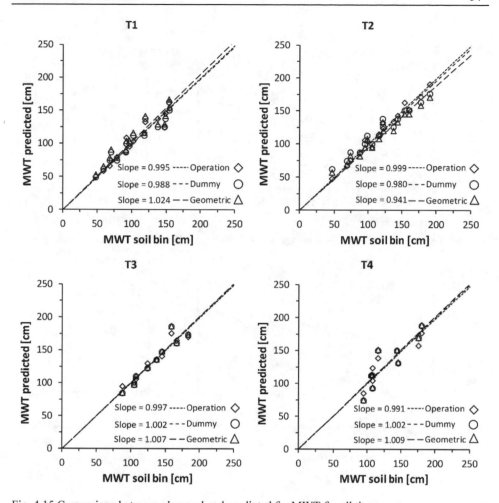

Fig. 4.15 Comparison between observed and predicted for MWT for all tines

4.2.2 Area

4.2.2.1 Effect of Speed and Depth on the Area of the Ridge

Effects of speed and depth on the Ar in cm² for all tines were plotted in Fig. 4.16 under soil bin conditions. From this figure and the regression model, it can be seen that a different manner is attained in this characteristic for every tine. The Ar for T1 and T2 increased linearly with the interaction between speed-depth (positive value of C_3) while Ar for T3 and T4 increased linearly with depth (positive value of C_2).

The maximum values of Ar were 423.6, 419.3, 384.7 and 322.7 cm² for T4, T2, T3 and T1, respectively, with variations of 1%, 9% and 24% between T4 and other tines, respectively.

Note: Based on the results, T4 and T2 can be used as a weed control tools and adding wings to T2 reduces the amount of Ar.

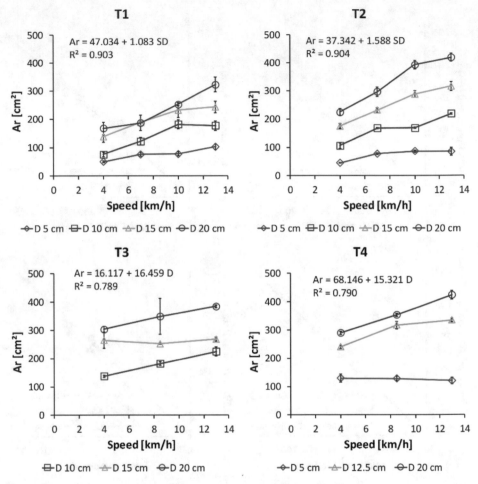

Fig. 4.16 Effects of speed and depth on Ar for all tines under soil bin conditions (mean ± SD)

The analysis of variance is presented in Table 4.16. It can be seen from this table that the effect of depth is of higher significance than speed on Ar for T1 ($P < 0.001$ and $P < 0.01$, respectively), T2 ($P < 0.000$ and $P < 0.001$, respectively), T3 ($P < 0.001$ and $P < 0.01$, respectively), and T4 ($P < 0.000$ and $P < 0.001$, respectively).

Table 4.16 ANOVA table for the area of the ridge

Source of variation	F – Value (Ar)			
	T1	T2	T3	T4
Speed	92.6*	309.9**	12.7*	118.2**
Depth	208.2**	1366.4***	102.9**	1242.8***
Speed × Depth	6.7*	28.3*	2.9	42.4*
R^2	0.968	0.994	0.931	0.994

* $P < 0.01$

** $P < 0.001$

*** $P < 0.000$

It can also be seen from this table that the interaction between speed-depth had a significant effect on Ar (P < 0.01) for all tines except T3.

4.2.2.2 Effect of Speed and Depth on the Af

Figure 4.17 shows the effect of speed and depth on the area of soil remaining in the furrow (Af) in cm² for all tines under soil bin conditions. From this figure and the regression model, it can be seen that a different manner is attained in this characteristic for every tine.

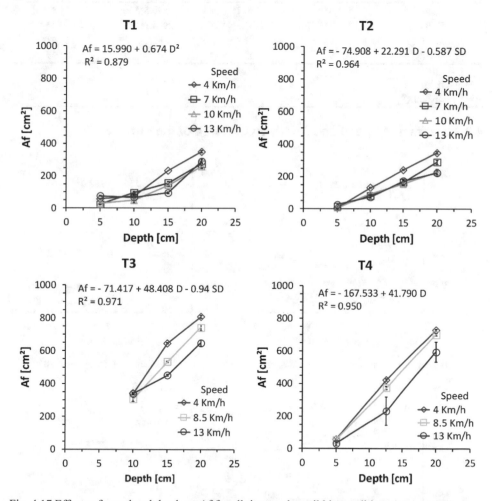

Fig. 4.17 Effects of speed and depth on Af for all tines under soil bin conditions (mean ± SD)

The Af for T1 increased quadratically with depth, while Af for T2 and T3 increased linearly with depth and with the interaction between the speed-depth term (Fig. 4.17). It can also be seen from this figure that the Af for T4 increased linearly with depth only. However, the maximum values of Af were 807.4, 726.7, 350.4 and 347.3 cm² at the deeper operating depth of 20 cm for T3, T4, T1 and T2, respectively, with variations of 10%, 56% and 57% between T3 and other tines, respectively. Note: According to the results, adding wings to T2 had a significant effect on Af.

The analysis of variance is presented in Table 4.17. It can be seen from this table that the effect of depth is of higher significance than speed on Af for all tines (P < 0.000 and P < 0.001, respectively). It can also be seen from this table that the interaction between speed-depth had a significant effect on Af (P < 0.01) for T1, T2, and T3 and (P < 0.001) for T4.

Table 4.17 ANOVA table for the area of soil remaining in the furrow

Source of variation	F – Value (Af)			
	T1	T2	T3	T4
Speed	146.6**	196.1**	256.5**	724.3**
Depth	2728.3***	3137.3***	2944.1***	19186.8***
Speed × Depth	59.0*	53.9*	68.5*	124.6**
R²	0.997	0.997	0.997	0.999

* P < 0.01
** P < 0.001
*** P < 0.000

4.2.2.3 Effect of Speed and Depth on the Area of the Furrow

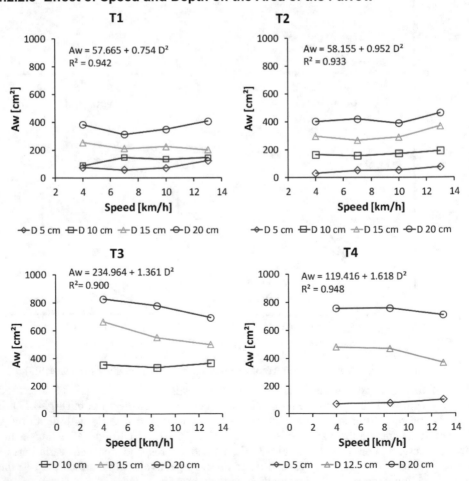

Fig. 4.18 Effects of speed and depth on the Aw for all tines under soil bin conditions

Effects of speed and depth on the Aw in cm² for all tines were plotted in Fig. 4.18 under soil bin conditions. From this figure and the regression model, it can be seen that a similar manner is attained in this characteristic for every tine. The Aw increased quadratically only with depth for all tines.

The maximum values of Aw were found to be 827, 760, 463 and 406 cm² for T3, T4, T2 and T1, respectively, at maximum operating depth of 20 cm, with variations of 8%, 44% and 51% between T3 and other tines, respectively.

Note: Based on the results, adding wings to T2 had a significant effect on Aw.

4.2.2.4 Three Regression Models to Predict the Area of the Furrow

Three regression models were used to predict Aw in the soil bin conditions (as mentioned above) for the previous section. The results are summarized in Table 4.18. It is obvious from this table that the coefficient of depth square C_5 is the main factor influencing Aw for each tine and for all regression models. Good values of the coefficient of determination were found ($R^2 > 90\%$) for all regression models.

From Table 4.18 it can be seen that the coefficient of depth square C_5 has similar values of 1.089 and 1.097 in the dummy and geometric model, respectively. This similarity can be attributed to the stable test environment done in specific soil type and conditions.

Table 4.18 Regression models and coefficients for the area of the furrow

Regression model	Tines				Regression coefficient						R^2
					C_0	C_1	C_2	C_3	C_4	C_5	
Operating condition	T1				57.665	n.s.	n.s.	n.s.	n.s.	0.754	0.942
	T2				58.155	n.s.	n.s.	n.s.	n.s.	0.952	0.933
	T3				234.964	n.s.	n.s.	n.s.	n.s.	1.361	0.900
	T4				119.416	n.s.	n.s.	n.s.	n.s.	1.618	0.948
		K			C_0	C_1	C_2	C_3	C_4	C_5	R^2
Dummy variable	T1	0.000									
	T2	37.638			- 5.115	n.s.	n.s.	n.s.	n.s.	1.089	0.914
	T3	305.768									
	T4	227.107									
Geometric variable		G_1	G_2	G_3	C_0	C_1	C_2	C_3	C_4	C_5	R^2
	T	7.613	n.s.	n.s.	- 64.078	n.s.	n.s.	n.s.	n.s.	1.097	0.910

n.s.: not significant

Table 4.18 shows that the coefficient K for a dummy regression model is equal to zero for T1 because it was set as the reference tine, and it increased to 37.638 and to 227.107 and then to 305.768 for T2, T4 and T3, respectively, with increasing tine width.

As expected from the previous regression, only the coefficient of tine width G_1 appeared in the geometric regression model with high values of 7.613 (Table 4.18).

4.2.2.5 Validation of the Regression Models for the Aw in the Field

In order to confirm the validity of the regression models obtained from the soil bin, the observed (field, geometric calculating) and the predicted (regression, laser scanner measuring) values of Aw for all tines are plotted in Fig. 4.19. It can be seen from this figure that the regression models of predictions had higher values of Aw than the observed field tests for T1 and T2. The reason is attributed to three factors: the first factor is related to the field condition being different from the soil bin in regard to the existence of stones, roots of the previous crop, and weeds that lead to an increase of the overall cohesion of the soil, which causes a reduction of soil rift rate. The second factor is due to Wf being one of the equation components for the calculation of Aw, and the third factor is associated with measurement accuracy.

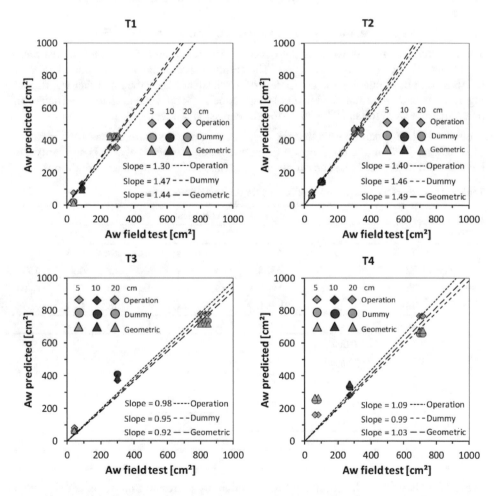

Fig. 4.19 Correlation between observed and predicted for Aw at different depths for all tines

Tine T1 shows a wide correlation between observed and predicted values of Aw (Fig. 4.19). From this graphic, it can be seen that the slopes of the best-fit line were 1.30 for operation condition, 1.44 for geometric, and 1.47 for the dummy. The regression equation model predicted the Aw of the geometric, dummy and operating condition regression with average absolute variations of 28%, 34% and 37%, respectively.

Figure 4.19 shows a wide range between observed and predicted values of Aw for T2 with slope 1.40 for operating condition, 1.46 for dummy and 1.49 for the geometric regression. The average absolute variations between observed and predicted values of Aw were found to be 43% for the dummy, 44% for the operating condition and 48% for the geometric regression model. An excellent acceptance between observed and predicted values of Aw was found (Fig. 4.19) for T3 with slope 0.98 for the operating condition, 0.95 for the dummy and 0.92 for the geometric regression model, with average absolute variation between observed and predicted values computed to be less than 5% for all regression models.

From the comparison in Fig. 4.19, it is clear that there is a very good accordance between observed and predicted values of Aw for T4 with slope 0.99 for the dummy and 1.03 for the geometric and 1.09 for the operating condition. The average absolute variation between observed and predicted values of Aw was calculated to be less than 18% for all regression models. Note: According to the results, T3 and T4 are best used for weed control and to cut the roots of the previous crop.

4.3 Correlation between Force and Soil Profile

4.3.1 Effects of Speed and Depth on the Specific Force

Effects of speed and depth on the specific force (Fs) in kN/m^2 are presented in Fig. 4.20 for all tines under the soil bin conditions. The Fs is defined as a ratio between force (Fd) and area (Aw). The minimum values of Fs are interesting here; it represents the least consuming force with highest soil cutting and distributing.

Figure 4.20 and the regression model show every tine having a different manner to achieve this characteristic.

The Fs increased with speed for all tines (positive value of C_1), while it decreased with depth for T2 and T4 (negative value of C_2) but not T1 (increasing). The minimum values for Fs were 21.7, 22.2, 23.9 and 25.6 kN/m^2 for T3, T1, T2 and T4, respectively, at a speed of 4 km/h and depth of 15, 5, 10 and 12.5 cm, respectively. The absolute variations were 2%, 10% and 18% between T3 and other tines, respectively. Therefore, it is preferred to use T2 with wings.

The Fs for T3 increased only with speed with a high value for the coefficient of determination R^2 equal to 0.915 and an acceptable value of 0.621 for T4, but with lower values of 0.488 and 0.312 for T1 and T2, respectively. Therefore, this leads to a good prediction for T3 and T4 and a weak prediction for T1 and T2 for Fs.

The analysis of variance is presented in Table 4.19. It can be seen from this table that the effect of speed and depth had the same influence on Fs for T1 and T4 (P < 0.01).

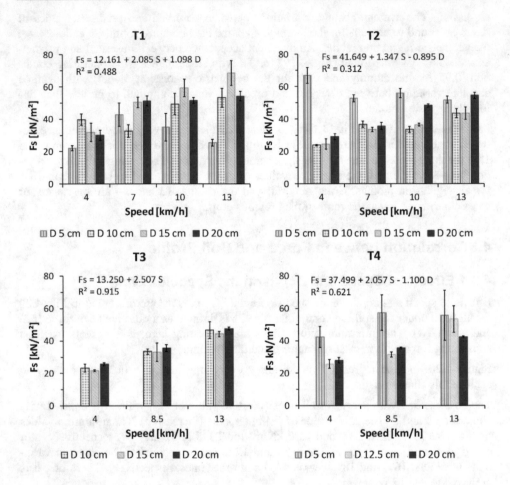

Fig. 4.20 Effects of speed and depth on Fs for all tines in the soil bin conditions (mean ± SD)

Table 4.19 ANOVA table for the specific force

Source of variation	F – Value (Fs)			
	T1	T2	T3	T4
Speed	37.4*	54.1*	133.8**	14.5*
Depth	38.6*	219.9***	2.9	14.0*
Speed × Depth	9.3*	33.0*	0.1	2.2
R^2	0.907	0.972	0.938	0.785

* $P < 0.01$
** $P < 0.001$
*** $P < 0.000$

It can also be seen from Table 4.19 that the interaction between speed-depth had a significant effect on Fs ($P < 0.01$) for T1 and T2, but not for T3 and T4.

4.3.2 Validation of the Regression Model for the Fs in the Field

To verify the accuracy of the regression model, which was obtained from the soil bin for the specific force (Fs), the predicted values (regression) are plotted against the observed values (field test) in Fig. 4.21 for all tines.

From this comparison, it can be seen that there is a good correlation between observed and predicted values of Fs for T3 and T4 and a weak correlation for T1 and T2.

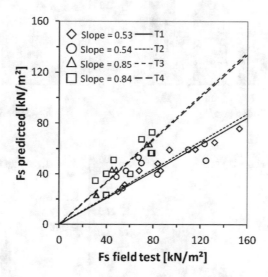

Fig. 4.21 Correlation between predicted (regression) and observed (field test) for Fs for all tines

The average absolute variations between predicted and observed values for Fs were found to be 45% for T1, 42% for T2, 16% for T3 and 18% for T4 and the slope of the best fit line was 0.53, 0.54, 0.85 and 0.84 for T1, T2, T3 and T4, respectively. This is attributed to two factors: the first factor is related to the area of the furrow Aw (a good prediction for T3 andT4) and the second factor is associated with the regression equation for Fs (a good prediction for T3 and T4).

4.3.3 Effects of Speed and Depth on the Specific Power

Figure 4.22 shows the effect of speed and depth on the specific power (Ps) in kW/m² for all tines under soil bin conditions. The Ps is defined as a power per unit area. The minimum values of Ps are the focus here; it represents the least consuming power with highest soil cutting and distributing.

From this figure and the regression model, it can be seen that a different manner is attained in this characteristic for every tine. The Ps increased with speed for all tines (positive value of C_1), while it decreased with depth for T2 and T4 (negative value of C_2), but not T1 (increasing). The minimum values for Ps were found to be 24.1, 24.7, 26.6 and 28.4 kW/m² for T3, T1, T2 and T4, respectively, at a speed of 4 km/h and depth 15, 5, 10 and 12.5 cm, respectively. Absolute variations were 3%, 10% and 18% between T3 and other tines, respectively.

Based on the results, it is preferred to use T2 with wings.

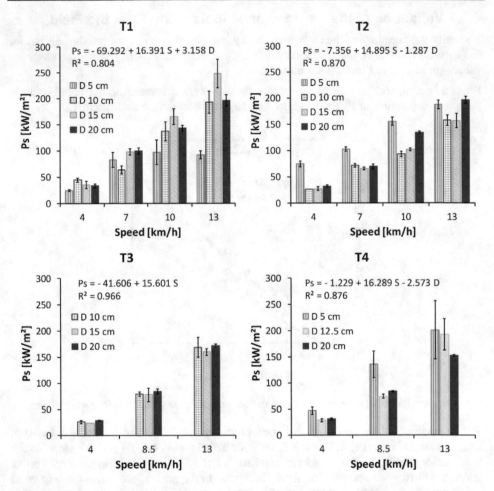

Fig. 4.22 Effects of speed and depth on Ps for all tines in the soil bin conditions (mean ± SD)

The analysis of variance is presented in Table 4.20. It can be seen from this table that the effect of speed is of higher significance than depth on Ps for T1, T2, and T3 (P < 0.000) and (P < 0.001) for T4.

Table 4.20 ANOVA table for the specific power

Source of variation	F – Value (Ps)			
	T1	T2	T3	T4
Speed	266.4***	1049.5***	645.3***	92.3**
Depth	44.7*	129.0**	1.9	6.9*
Speed × Depth	14.4*	11.3*	0.3	1.8
R^2	0.971	0.991	0.989	0.920

* P < 0.01
** P < 0.001
*** P < 0.000

It can also be seen from Table 4.20 that the interaction between speed-depth had significant effects on Ps (P < 0.01) for T1 and T2, but not T3 and T4.

4.3.4 Validation of the Regression Model for the Ps in the Field

To confirm the veracity of the regression model, which was obtained from the soil bin for the specific power (Ps), the predicted values (regression) are plotted against the observed values (field test) in Fig. 4.23 for all tines.

From this evaluation a good correlation for T3 and T4, acceptable for T2, and weak for T1 was found between observed and predicted values of Ps.

The slope of the best-fit line was 0.56 for T1, 0.63 for T2, 0.92 for T3 and 0.78 for T4 and the average absolute variations between predicted and observed values for Ps were 45%, 36%, 12% and 19% for T1, T2, T3 and T4, respectively.

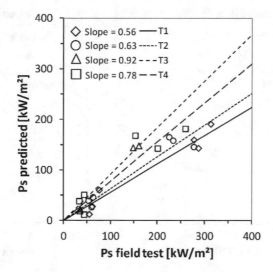

Fig. 4.23 Correlation between predicted (regression) and observed (field test) for Ps for all tines

Note: The speed of 20 km/h was neglected from the comparison for two reasons.

- The first reason is due to Ps being highly affected by speed.

- The second reason is related to the speed of 20 km/h being beyond the scope of prediction.

4.4 Soil Loosening Percentage

4.4.1 Effects of Speed and Depth on the SLa%

Effects of speed and depth on the soil loosening percentage above the soil surface (SLa%) are presented in Fig. 4.24 for all tines under the soil bin conditions. the SLa% refers to the amount of soil loosened above the original soil surface and it is a ratio between Ar and At.

Figure 4.24 and the regression model show a different manner for every tine to achieve this characteristic.

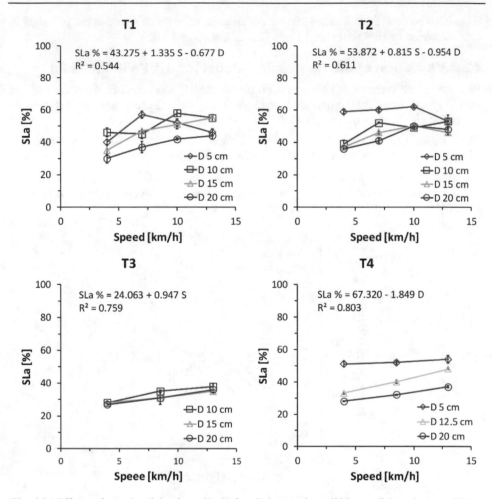

Fig. 4.24 Effects of speed and depth on SLa% for all tines under soil bin conditions (mean ± SD)

The SLa% increased with speed for all tines (positive value of C_1), except T4 (no effects); while it decreased with depth for all tines (negative value of C_2), except T3 (no effects). The maximum values for SLa% were 62%, 58%, 54% and 38% for T2, T1, T4 and T3, respectively, with absolute variations of 6%, 12% and 39% between T2 and other tines, respectively. Therefore, adding wings to T2 had a significant effect on the SLa% (decreasing).

4.4.2 Effects of Speed and Depth on the SLu%

Effects of speed and depth on the soil loosening percentage under the soil surface (SLu%) are presented in Fig. 4.25 for all tines under the soil bin conditions. The SLu% refers to the amount of soil loosened under the original soil surface or so-called furrow backfill (loosened soil remaining in the furrow) and it is a ratio between Af and At.

Figure 4.25 and the regression model show a different behavior for every tine regarding this characteristic.

The SLu% decreased with speed for all tines (negative value of C_1), while it increased with depth for all tines (positive value of C_2) except T3 (no effects).

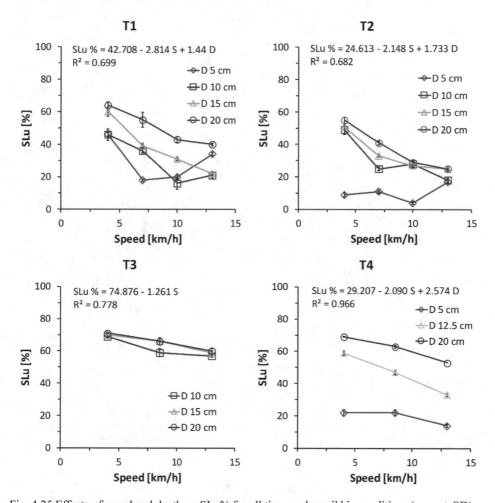

Fig. 4.25 Effects of speed and depth on SLu% for all tines under soil bin conditions (mean ± SD)

The maximum values for SLu% were 71%, 69%, 64% and 55% for T3, T4, T1 and T2, respectively, with absolute variations of 3%, 10% and 23% between T3 and the other tines, respectively. Therefore, adding wings to T2 had a significant effect on the SLu% (increasing).

5 Comparison with other Models and Results

5.1 Force

5.1.1 Horizontal Force According to Gorjatschkin

Gorjatschkin, (1927) proposed an empirical formula for predicting the horizontal force acting on the moldboard plow, as presented in Eq. (5.1).

$$Fh = Wt \times d \times (G + \omega \times S^2) \tag{5.1}$$

Where G is the coefficient of static resistance in (kN/m²), and ω is the constant of dynamic draft in (kN s²/m⁴). The problem of Gorjatschkin's formula is in the magnitude of these two factors.

Soucek & Pippig (1990) suggested a value of G for a sandy loam soil between 40...60 kN/m². While ω is between 3...9 kN s²/m⁴.

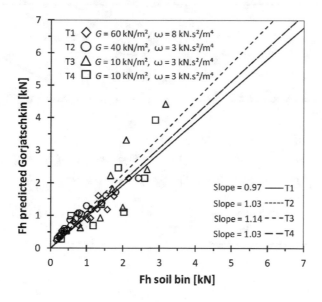

Fig. 5.1 Correlation between Fh soil bin and Fh predicted Gorjatschkin's formula for all tines

The measured Fh from soil bin and the predicted Fh based on Gorjatschkin's formula for all tines are plotted in Fig. 5.1. From this figure, a good correlation between measured and predicted Fh for all tines can be seen. The average absolute variations were 1%, 2%, 9% and 11% for T4, T1, T3 and T2, respectively. With slopes were 0.97 for T1, 1.14 for T3 and 1.03 for both T2 and T4.

Figure 5.1 indicates that the coefficient G was within the range of presumption from Soucek & Pippig for T1 and T2, while it was under the presumption for T3 and T4. Fig. 5.1 also shows that the magnitude of coefficient ω was very high for T1, while it was very low for T2, T3 and T4. Therefore, the advantage and disadvantage of using Gorjatschkin's formula can be summarized as follows.

© Springer-Verlag GmbH Germany, part of Springer Nature 2019
A. K. A. Al-Neama, *Evaluation of performance of selected tillage tines regarding quality of work*, Fortschritte Naturstofftechnik, https://doi.org/10.1007/978-3-662-57744-8_5

The advantage is the formula is very simple as it has only two coefficients, while the disadvantage can be attributed to two reasons. There is a wide range of magnitude for these coefficients and there is no base to choose the magnitude of these coefficients.

5.1.2 Fh and Fv Forces According to Mckyes and Ali's Model

McKyes & Ali's model to predict the horizontal and vertical forces acting on the narrow flat blades are presented in Eq. (2.21) and (2.24), respectively, after substituting Ft with a dynamic term related to the speed (see Eq. (2.25) section 2.2.2.2.3). A MATLAB based computer program was used in order to compute these N factors.

Figure 5.2 presents the correlation between regression equations for Fh and Fv obtained from the soil bin and McKyes & Ali's model for all tines at different depths with three chosen levels of rake angle (30°, 50° and 70°) at constant speed 10 km/h.

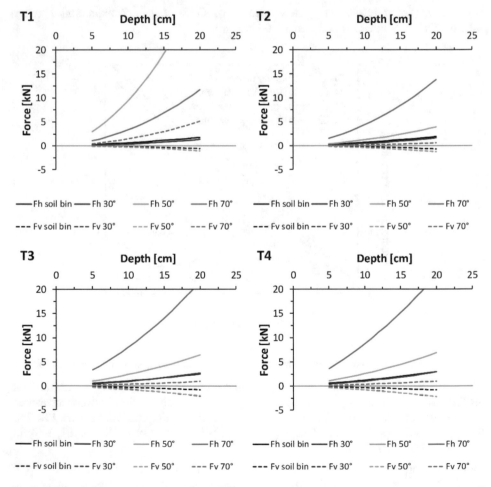

Fig. 5.2 Correlation between regression equation and McKyes & Ali's model for Fh and Fv for all tines at different depths and rake angles 30°, 50° and 70° at constant speed 10 km/h

From this figure, a good correlation between regression and predicted values based on McKyes & Ali's model for Fh can be seen for all tines at rake angle 30° with the range of depths between 5 to 20 cm (the black and red solid lines, respectively).

The average absolute variation between regression equation and McKyes & Ali's model at rake angle 30° and speed of 10 km/h were 33%, 23%, 9% and 1% for T1, T2, T4 and T3, respectively. From this figure a good correlation between regression and predicted values based on McKyes & Ali's model for Fv can be seen for T1 and T2 with rake angle 30° and 50° at the range of depths between 5 to 20 cm (the black, red, and green dotted lines, respectively). T3 and T4 showed a good correlation between regressions and predicted for Fv with rake angle 30° and 50° at shallow depths only (note the Fv value at rake angle 70° is positive). This is due to McKyes & Ali's use of a narrow blade.

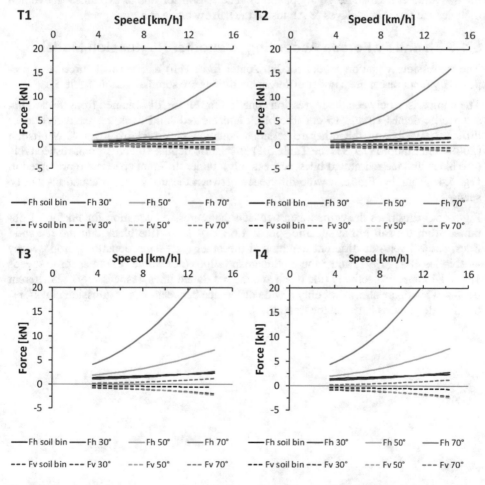

Fig. 5.3 Correlation between regression equation and McKyes & Ali's model for Fh and Fv for all tines at different speeds and rake angles 30°, 50° and 70° at constant depth 15 cm

Figure 5.3 illustrates the correlation between regression equations for Fh and Fv obtained from the soil bin and McKyes & Ali's model for all tines at different speeds with three selected magnitudes of rake angle (30°, 50° and 70°) at constant depth 15 cm. From this figure, a good correlation between regression and predicted values based on McKyes & Ali's model for Fh can be seen for all tines with rake angle 30° at the range of speeds between 4 to 14 km/h (the black and red solid line, respectively).

The average absolute variations between regression equation and McKyes & Ali's model at rake angle 30° and depth of 15 cm were 32%, 20%, 6% and 5% for T1, T2, T3 and T4, respectively.

From this figure, a good correlation between regression and predicted values based on McKyes & Ali's model for Fv can be seen for T1 and T2 with rake angles 30° and 50° at the range of speeds between 4 and 14 km/h (the black, red, and green dotted lines, respectively). While T3 and T4 showed a good correlation between regressions and prediction for Fv with rake angles 30° and 50° at lower speeds between 4 to 6 km/h only (note the Fv value at rake angle 70° is positive). The reason for this is explained above and includes the fact that McKyes & Ali tested the narrow blade at a low speed.

5.1.3 Fh and Fv Forces According to another Regression Model

The regressions equation model for horizontal force (Fh) and vertical force (Fv) proposed by other researchers under different conditions are summarized in Table 5.1.

The comparison between the regression equation for Fh for T1 obtained from the soil bin at specific depths 10 and 15 cm (the black and the red solid lines, respectively) under different speeds and the other regressions equation proposed by Rosa & Wulfsohn (2008) and Stafford (1979) (see Table 5.1) at specific depths 10 and 15 cm, respectively (the black and the red dotted lines, respectively), under different speeds are presented in Fig. 5.4. From this figure, a wide difference between these regression equations can be seen.

Figure 5.5 illustrates the comparison between the regression equation for Fh for T1 obtained from the soil bin at specific speeds 1.6 and 4 km/h (the black and the red solid lines, respectively) at different depths and the other regression equations proposed by Desbiolles et al. (1997) and Sahu & Raheman (2006) (see Table 5.1) at specific speeds 1.6 and 4 km/h, respectively (the black and the red dotted lines, respectively), at different depths. This figure shows not only a wide difference between these regression equations, but also differences between the trend lines.

Table 5.1 Regression equations for Fh and Fv proposed by other researchers under different conditions

Name and Year	Tines Used	Soil Type	Soil Physical Properties	Operation Condition*	Regression Equations	Regression Coefficients
Stafford 1979	plane rigid tine $Wt = 4$ cm $\alpha = 45°$	Sandy clay loam Soil bin	$MC_d = 12.6 \% \pm 0.3$ $\rho_d = 1490$ kg/m³ ± 30 $\phi = 32.5° \pm 2.1$	$S = 0.5$ to 5.5 m/s $d = 15$ cm	$Fh = a + bS + cS^2$ Fh [kN]	$a = 0.264$ $b = -0.06$ $c = 0.06$
Rosa & Wulfsohn 2008	vertical rigid tines $Wt = 9.4$ mm	Silty clay loam Soil bin	$MC_d = 12 \% \pm 0.5$ $\rho_d = 1180$ kg/m³ ± 4	$S = 0.5$ to 10 m/s $d = 10$ cm	$Fh = a + bS$ Fh [N]	$a = 102$ $b = 123$
Desbiolles et al. 1997	Curved chisel tine $Wt = 7.5$ cm	Sandy loam Soil bin	$MC_d = 9.8 \%$ $\rho_d = 1490$ kg/m³ $\phi = 23.3° \pm 2$	$S = 1.6$ km/h $d = 0.10$ to 0.30 m	$Fh = a + bd + cd^2$ Fh [kN]	$a = 1.858$ $b = -7.756$ $c = 17.364$
Sahu & Raheman 2006	Standard tine $Wt = 7.5$ cm	Sandy clay loam Soil bin	$MC_d = 10.3 \%$ $\rho_d = 1170$ kg/m³ ± 20 $\phi = 22°$	$S = 1.2$ to 4.2 km/h $d = 0.05$ to 0.10 m	$Fh = a + bd + cdS$ Fh [N]	$a = 0.0$ $b = 726$ $c = 241$
Onwualu & Watts 1998	Wide plane tine $Wt = 25.4$ cm $\alpha = 45°$	Silty sand Soil bin	$MC_d = 14 \%$ $\rho_d = 1500$ kg/m³ $\phi = 30°$	$S = 0.25$ to 2.0 m/s $d = 15$ cm	$Fh = a + bS + cS^2$ Fh [N]	$a = 721$ $b = 280$ $c = 9$
Onwualu & Watts 1998	Wide plane tine $Wt = 25.4$ cm $\alpha = 45°$	Silty sand Soil bin	$MC_d = 14 \%$ $\rho_d = 1500$ kg/m³ $\phi = 30°$	$S = 0.25$ to 2.0 m/s $d = 15$ cm	$Fv = a + bS + cS^2$ Fv [N]	$a = 141$ $b = 261$ $c = -65$

*Used the original unit for substitution in the regression equations

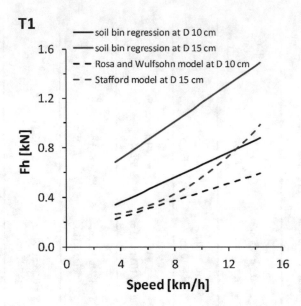

Fig. 5.4 Compression between the soil bin regression equation and the other regression equations proposed by Rosa & Wulfsohn (2008) and Stafford (1979) at depths 10 and 15, respectively, for T1 under different speeds for Fh

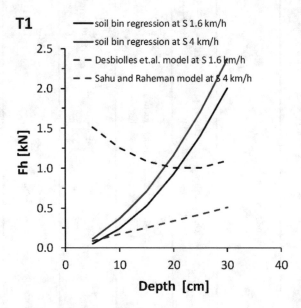

Fig. 5.5 Compression between the soil bin regression equation and the other regression equations proposed by Desbiolles et al. (1997) and Sahu & Raheman (2006) at speeds 1.6 and 4 km/h, respectively, for T1 under different depths for Fh

Figure 5.6 shows the compression between the regression equation for T3 for both Fh and Fv (the black and the red solid lines, respectively) obtained from the soil bin at a specific depth of 15 cm under different speeds and the other regression equations proposed by Onwualu & Watts for a wide plane tine for both Fh and Fv (the black and the red dotted lines, respectively) (see Table 5.1).

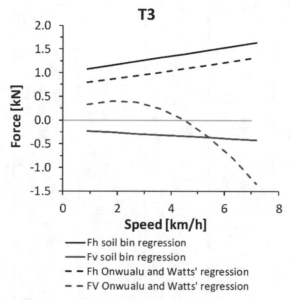

Fig. 5.6 Compression between the soil bin regression equation and Onwualu & Watts' (1998) regression equation at a depth of 15 cm for T3 under different speeds for Fh and Fv

From this figure, a good correlation between the soil bin regression and Onwualu & Watts' regression (the solid and dotted black line, respectively) for Fh can be seen. The average absolute variation between these regressions equation was 22.6%. A wide difference between the soil bin regression and Onwualu & Watts' regression (the solid and dotted red line, respectively) for Fv can be seen from Fig. 5.6, not only by their magnitudes but also by their trend line.

5.2 Soil Profile

5.2.1 Soil Profile Parameters According to Other Regression Models or to the Original Data

The soil profile parameters according to other regression models or to the original data were obtained from the soil bin or field test by using standard tines or a simple plane blade under different operation conditions as summarized in Table 5.2. From this table it can be seen that Stafford (1979) (first row) found that Aw increased linearly with speed at constant depth. Different results have been obtained from the soil bin for all tines T1, T2, T3, and T4 at different operation conditions, where Aw increased quadratically with depth only.

Table 5.2 Regression equations for soil profile parameters or the original data proposed by other researchers under different conditions

Name and Year	Tines Used	Soil Type	Soil Physical Properties	Operation Condition*	Regression Equations	Regression Coefficients
Stafford 1979	plane rigid tine $Wt = 4$ cm $\alpha = 45°$	Sandy clay loam Soil bin	$MC_d = 12.6\,\% \pm 0.3$ $\rho_d = 1490$ kg/m³ ± 30 $\phi = 32.5° \pm 2.1$	$S = 0.5$ to 5.5 m/s $d = 15$ cm (constant)	$Aw = a + bS$ Aw [cm²]	$a = 207$ $b = 20.6$
Willatt & Willis 1965	curved and vertical plane tines $Wt = 2$ in.	Soil range between sandy loam to clay Field	$MC_d = 24\,\%$	$S = 4$ to 5 ft./s $d = 2$ to 7 in.	$Wf = a* d + Wt$ $Aw = b*d^2 + Wt*d$ Wf [in.], Aw [in.²]	$a = 2.42$ $b = 1.03$
Liu & Kushwaha 2005	Standard wide sweep tine $Wt = 32.5$ cm	Sandy loam Soil bin	$MC_d = 12\,\%$ $\rho_d = 1280$	$S = 5, 7.5, 10$ km/h $d = 10$ cm (constant)	Original data for RRD [mm] MWT [mm]	-------
Manuwa 2009	vertical plane tines $Wt = 5$ cm $\alpha = 90°$	Sandy clay loam Soil bin	$MC_d = 11.5\,\%$ $\rho_d = 1520$ kg/m³ ± 10 $\phi = 29.6° \pm 1.2$	$S = 3.6$ km/h (constant) $d = 10$ and 15 cm	Original data for RRD [cm], Wf [cm] MWT [cm]	-------
McKyes & Desir 1984	Flat blade $Wt = 6.3, 12.5$ cm $\alpha = 35°$	Sandy clay loam Field	$MC_d = 21.9\,\% \pm 1.4$ $\rho_d = 1450$ kg/m³ ± 12 $\phi = 36.0° \pm 3.1$	$S = 5$ km/h (constant) $d = 15$ and 25 cm	Original data for Aw [m²] and SLa [%]	-------

*Used the original unit to substitute into the regression equations (in. = inch, ft. = feet)

5.2.2 Width and Area of the Furrow According to the Willatt & Willis Model

Willatt & Willis (1965) proved that Wf increased linearly with increased operating depth while Aw increased quadratically with increased operating depth under different operation conditions (see Table 5.2). Identical results have been found from the soil bin for all tines. The measured values of Wf obtained from the soil bin for T1 and T2 by using the laser scanner and the Willatt & Willis prediction equations are plotted in Fig. 5.7 at different depths. Note that the comparisons were done according to matching tine widths (excluding T3 and T4).

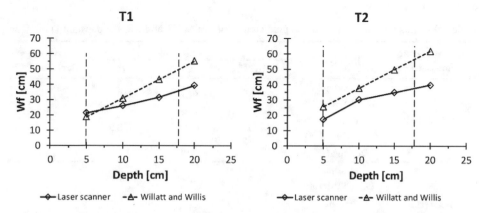

Fig. 5.7 Correlation between measured and predicted according to Willatt & Willis' model for Wf

From this figure, it can be seen that there is a good correlation between measured and predicted value of Wf for T1 and T2 at shallow depths of 5 and 10 cm. The range of depths tested by Willatt & Willis is represented by the two vertical dotted lines in Fig. 5.7. The absolute variations were 13% and 18% for T1 and 26% and 20% for T2 at depths 5 and 10, respectively.

The measured value of Aw (laser scanner) is plotted against the predicted value based on Willatt & Willis's model in Fig. 5.8.

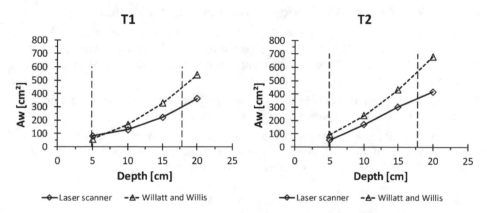

Fig. 5.8 Correlation between measured and predicted based on Willatt & Willis' model for Aw

From this figure, it can be seen that there is a better agreement between measured and predicted value of Aw for T1 and T2 at the shallow depths of 5 and 10 cm than for 15 and 20 cm. The range of depths tested by Willatt & Willis is indicated by two vertical dashed lines in Fig. 5.8. The absolute variations were 22% and 27% for T1 and 38% and 26% for T2 at depths 5 and 10, respectively.

5.2.3 MWT and RRD Original Data Based on Liu & Kushwaha

Comparison of measured (Liu & Kushwaha data) and predicted (regression equation from soil bin) of MWT and RRD are presented in Table 5.3 at a constant depth of 10 cm with varying speed. Note that the comparison was done with T4 because Liu & Kushwaha used a sweep tine.

Table 5.3 Comparison of measured and predicted values of MWT and RRD at constant D of 10 cm

Soil disturbance parameters	Data type	Speed [km/h]			Average Deviation [%]
		5	7.5	10	
MWT [cm]	Measured Liu & Kushwaha	77.6	94.4	112.8	
	Predicted Regression (T4)	103.8	118.6	133.5	
	Deviation [%]	33.8	20.4	15.5	23.2
RRD [cm]	Measured Liu & Kushwaha	34.4	42.8	50.0	
	Predicted Regression (T4)	51.5	55.8	60.2	
	Deviation [%]	33.2	23.3	16.9	24.5

From this table, it can be seen that measured and predicted values of MWT and RRD share the same behavior, they increase with increased speed.

From this table, it can also be seen that there is a good agreement between measured and predicted values at a speed of 10 km/h for MWT and RRD with absolute variations of 15.5% and 16.9%, respectively.

5.2.4 MWT, RRD, and Wf Original Data Based on Manuwa

A comparison of measured (Manuwa data) and predicted (regression equation from soil bin) values of MWT, RRD, and Wf is presented in Table 5.4 at a constant speed of 3.6 km/h with varying depth. Note that the comparison was done with T1 because Manuwa used a 5 cm vertical plane tine.

From this table, it can be seen that there is a similar behavior for measured and predicted values of MWT and Wf, they increase with increased depth.

From this table, it can also be seen that there is a wide difference between measured and predicted values of the soil profile parameters, with average absolute variations of 43.1%, 40.7% and 27.0% for MWT, RRD and Wf, respectively.

Table 5.4 Comparison of measured and predicted values of MWT, RRD, and Wf at constant speed of 3.6 km/h

Soil disturbance parameters	Data type	Depth [cm]		Average Deviation [%]
		10	15	
MWT [cm]	Measured Manuwa	27.5	33.0	
	Predicted Regression (T1)	45.3	62.2	
	Deviation [%]	39.3	46.9	43.1
RRD [cm]	Measured Manuwa	13.5	16.5	
	Predicted Regression (T1)	25.3	25.3	
	Deviation [%]	46.6	34.8	40.7
Wf [cm]	Measured Manuwa	20.0	23.0	
	Predicted Regression (T1)	26.6	32.5	
	Deviation [%]	24.8	29.2	27.0

5.2.5 Furrow Area and Soil Loosening Percentage above the Soil Surface Based on McKyes & Desir's Original Data

A comparison of measured (McKyes & Desir data) and predicted (regression equation from soil bin) values of Aw and SLa % is plotted in Fig. 5.9 and Fig. 5.10, respectively, at a constant speed of 5 km/h with varying depth (see Table 5.2).

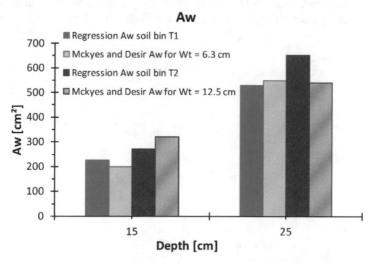

Fig. 5.9 Comparison of measured and predicted values of Aw at speed 5 km/h under varying depths

Note that the bar with full color blue or red is for T1 and T2, respectively, while the bars with dashed color blue or red are for narrow and wide flat blades 6.3 and 12.5 cm, respectively. Note also that the comparison was done according to the matching of the tine width T1 with 6.3 cm and T2 with 12.5 cm. From Fig. 5.9, it can be seen that the measured and predicted Aw increased with increasing depth. A good correlation between measured and predicted Aw was found, with absolute variations of 12% and 4% for T1 and 16% and 17% for T2 at depths 15 and 25 cm, respectively.

Figure 5.10 shows that the measured and predicted values of SLa% decreased with increasing depth.

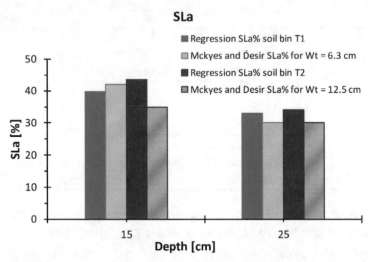

Fig. 5.10 Comparison of measured and predicted values of SLa% at a speed of 5 km/h under varying depths

A good agreement was found between measured and predicted Sla%, with absolute variations of 6% and 9% for T1 and 20% and 12% for T2 at a depth of 15 and 25 cm, respectively.

6 Draft Force Prediction Approach

6.1 Draft Force Prediction Model

This chapter explains a model to predict the draft force acting on varying standard single tines by using principles of soil mechanics and soil profile evaluation as presented in Eq. (6.1) (Al-Neama & Herlitzius, 2017).

$$Fd = Fp + Fg + Fi + Fc \tag{6.1}$$

Where Fp is the penetration force in kN and is measured by using a penetrometer (see Fig. 3.4).

Note: that the friction force (Ff), the adhesion force (Fca) and the surcharge pressure vertically acting on the soil surface (q) were neglected.

Fg is the gravitational force in kN and is given by Eq. (6.2).

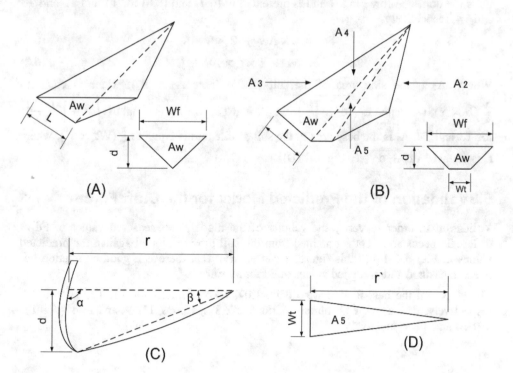

Fig. 6.1 Cross section of soil profile predictions

$$Fg = \rho_w \times V \times g \tag{6.2}$$

Where: V is the soil volume cut by the tine in m³. Figure 6.1 specifies the shape of the swept volume. For T1 and T2, the shape of the soil profile is equal to a triangle-based pyramid (see Fig. 6.1 (A)), while it is equal to a trapezoidal-based pyramid (see Fig. 6.1 (B)) for T3 and T4. Therefore, the swept volume is given by Eq. (6.3).

© Springer-Verlag GmbH Germany, part of Springer Nature 2019
A. K. A. Al-Neama, *Evaluation of performance of selected tillage tines regarding quality of work*, Fortschritte Naturstofftechnik, https://doi.org/10.1007/978-3-662-57744-8_6

$$V = \frac{1}{3}(Aw \times r) \qquad (6.3)$$

Aw is the area of the furrow; it is measured by laser scanner (see Fig. 3.8) and calculated by using Eq. (3.7). r is the rupture distance and is calculated by Eq. (6.4) as proposed by McKyes & Ali (1977).

$$r = d \times (\cot \alpha + \cot \beta) \qquad (6.4)$$

Where: α is the rake angle, which is equal to 45° for a standard tine (Desbiolles et. al. 1997), and β is the soil failure angle, which is obtained from passive earth pressure theory Eq. (6.5).

$$\beta = \frac{\pi}{4} - \frac{\phi}{2} \qquad (6.5)$$

Fi is the inertial force in kN and is given by Eq. (6.6).

$$F_i = \rho_w \times S^2 \times Aw \qquad (6.6)$$

Fc is the cohesion force in kN and is presented in (6.7) and (6.8) for T1 and T2 and T3 and T4, respectively.

$$Fc = C \times Aw + 2 \times C \times A_2 \qquad (6.7)$$

$$Fc = C \times Aw + 2 \times C \times A_2 + C \times A_5 \qquad (6.8)$$

Where: A_2 is the side area of the soil profile (note $A_2 = A_3$), which is equal to $\frac{1}{2}(L \times r)$, L is equal to $\sqrt{d^2 + \left[\frac{wf}{2}\right]^2}$ for T1 and T2, while it is equal to $\sqrt{d^2 + \left[\frac{wf-Wt}{2}\right]^2}$ for T3 and T4. A_5 is the base area of the soil profile and is equal to $\frac{1}{2}(Wt \times r`)$, where r` is equal to $\sqrt{r^2 + d^2}$ (see Fig. 6.1 (D)).

6.2 Validation of the Predicted Model for the Draft Force

Validation In order to verify the validity of the model, the measured values of Fd at different speeds and depths obtained from the soil bin are plotted against the predicted values in Fig. 6.2. From this figure, it can be seen that there is a good correlation between measured and predicted values of Fd for all tines.

The slopes of the best-fit line were 0.99, 1.07, 1.12 and 1.13 for T1, T2, T4 and T3, respectively, with average absolute variations of 2.8%, 10.4%, 11.5% and 13.8% for T1, T2, T4 and T3, respectively.

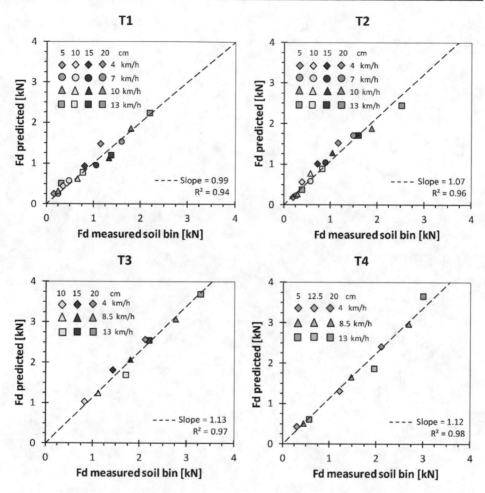

Fig. 6.2 Correlation between measured and predicted Fd for all tines at varying speeds and depths

7 Conclusions

7.1 Forces

- The horizontal, vertical, and draft forces (total force) increase linearly with the speed-depth interaction term, while they increased quadratically with the depth for all tines.

- The depth affects forces (horizontal, vertical, and draft force) more significantly in comparison to speed for all tines ($P < 0.000$ and $P < 0.001$, respectively).

- Double Heart with Wing tine registers the maximum value of all forces (horizontal, vertical, and draft force) than the other tines at higher operation conditions of speed and depth.

- Vertical force magnitude is negative at a depth of 5 cm and it changes to a positive magnitude at depths deeper than that for Heavy Duty and Double Heart tines.

- Empirical regression models for prediction forces (horizontal, vertical, and draft force) acting on standard single tines are developed by adding new coefficients related to tine geometry or by using a dummy variable. With these models, it is possible to calculate forces for a wide range of tine shapes using parameters of tool geometry under various operating conditions of speed and depth. Only the coefficient of tine width appeared in the geometric regression model, while the dummy coefficient increased with increasing tine width.

- Statistically it can be assumed that the horizontal force is equal to the draft force for all tines.

- A predicted draft force model acting on varying standard single tines based on principles of soil mechanics and soil profile valuation is verified.

- It is very difficult to use Gorjatschkin's formula to predict the horizontal force for all tines. Because of there is a wide range of magnitude for Gorjatschkin's coefficients and there is no base to choose the magnitude of these coefficients.

- Analytical model based on Mckyes and Ali's model show a good prediction for the horizontal force by substituting the rake angle of 30° in the model for all tines, while it is equal to 30° or 50° for the vertical force for Heavy Duty and Double Heart tines only.

- It is very difficult to predict vertical force by using another regression equation under different operation conditions for all tines, while some regression equations under different operation conditions are found to predict horizontal and draft force (at the same tine width).

7.2 Soil Profile Parameters

- Furrow height is highly affected by tine shape. It is affected by speed, by speed and depth, and by depth for Heavy Duty, Double Heart, and Duck Foot, respectively. It is not affected by speed or depth for Double Heart with Wing. Based on those results, Double Heart can be used as a furrow opener in seeding equipment.

© Springer-Verlag GmbH Germany, part of Springer Nature 2019
A. K. A. Al-Neama, *Evaluation of performance of selected tillage tines regarding quality of work*, Fortschritte Naturstofftechnik, https://doi.org/10.1007/978-3-662-57744-8_7

- Ridge height is not affected by tine shape. It increases only with depth for all tines. Based on this result, adding wings does not affect this parameter. In addition, coefficients of width, length, and tine angle in the geometric regression equation did not have significant effects on this parameter.

- Furrow width increases only with depth for all tines. Double Heart with Wing tine achieves maximum value (the widest). The dummy coefficient increases with increasing tine width in the dummy regression model. Only the coefficient of tine width appears in the geometric regression model. Willatt & Willis' equation can be used to predict furrow width for Heavy Duty and Double Heart tines at shallow depths.

- Ridge to Ridge Distance increases only with speed for all tines. Adding wings does not affect its value.

- Maximum width of soil throw increased with increasing speed and depth for all tines. The highest value is attained by Double Heart tine. Only the coefficient of length of the tine appears in the geometric regression model with a negative value, while the dummy coefficient decreases with increased tine length in the dummy regression model.

- Ridge area is highly affected by tine shape. It increases linearly with speed-depth interaction for Heavy Duty and Double Heart, while it increases linearly with depth only for Double Heart with Wing and Duck Foot. Duck Foot and Double Heart reach its maximum value. Therefore, these two tines are used in weed control. Furthermore, adding wings reduces the amount of this parameter.

- The area of soil remaining in the furrow is highly affected by tine shape. It increases quadratically with depth for Heavy Duty and linearly with depth for Duck Foot, while it increases with depth and decreases with speed-depth interaction for Double Heart and Double Heart with Wing. The Double Heart with Wing tine at deeper operating depths achieves the maximum value of this parameter. Furthermore, adding wings increased the amount of this parameter significantly.

- Furrow area increases quadratically only with depth for all tines. Double Heart with Wing tine achieves the maximum value of this parameter at deeper operating depths. Adding wings had a significant increase on the amount of this parameter. In addition, the dummy coefficient increases with increasing tine width in the dummy regression model, while the coefficient of tine width appears in the geometric regression model. Willatt & Willis' equation to predict furrow area can be used for Heavy Duty and Double Heart tines at shallow depths.

7.3 Soil Loosening Percentage

- Soil loosening percentage above the soil surface (original soil) is highly affected by tine shape. It increases with speed for all tines except Duck Foot (no effects), while it decreases with depth for all tines except Double Heart with Wing (no effects). Double Heart tine achieves the maximum value of this characteristic. Therefore, it highly decreases by adding wings.

- Soil loosening percentage under the soil surface (original soil) is highly affected by adding wings. It decreases with speed for all tines, while it increases with depth for all tines except Double Heart with Wing (no effects). Thus, Double Heart with Wing tine shows the maximum magnitude for this parameter.

7.4 Specific Force and Specific Power

• Specific force and specific power are highly affected by tine shape. They increase linearly with speed for all tines, while they decrease linearly with depth for Double Heart and Duck Foot, but increase with Heavy Duty. Double Heart with Wing tine achieves the minimum values of these parameters at depth 15 cm with slower speeds.

7.5 Summary

The characteristic behaviors of all studied parameters affected by speed and depth are summarized in Fig. 7.1.

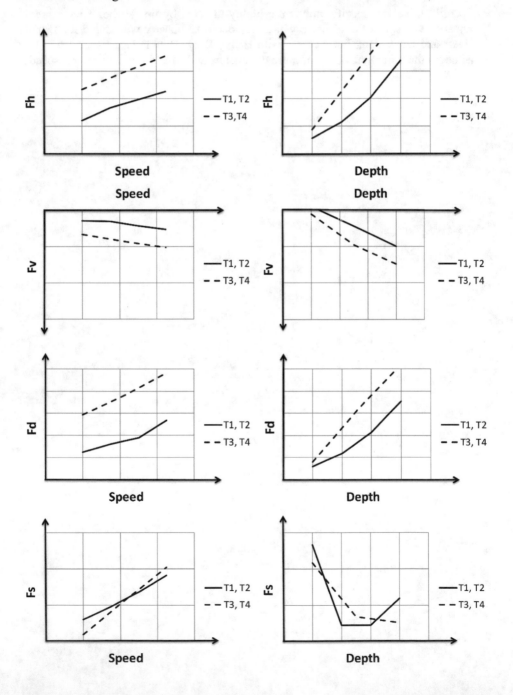

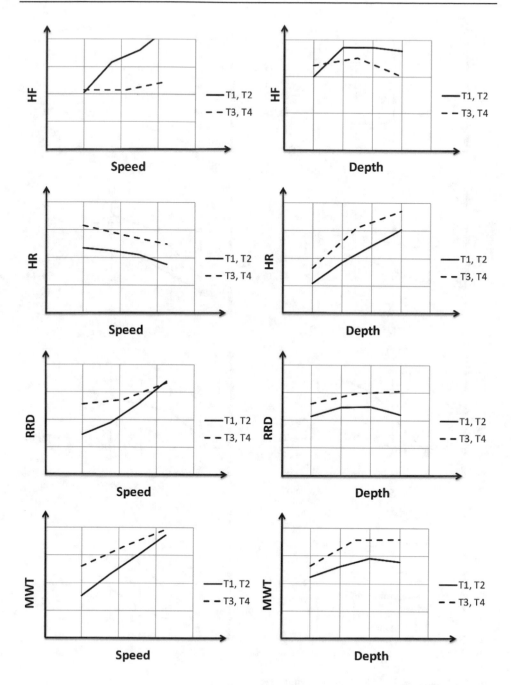

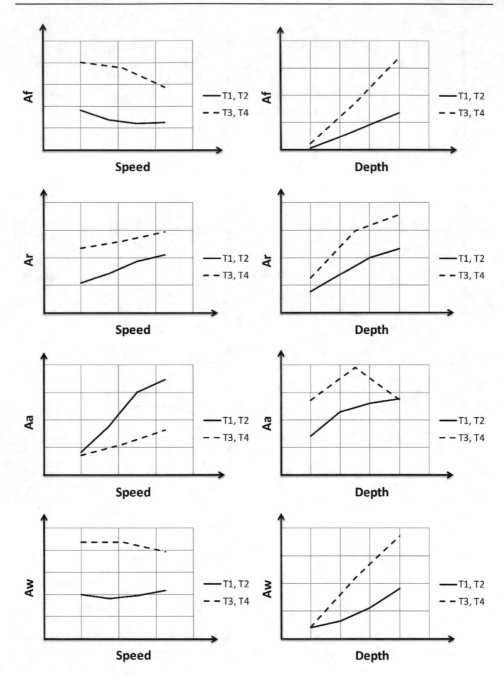

Fig. 7.1 Characteristic behaviors of all studied parameters affected by speed and depth

References

Al-Neama, A., & Herlitzius, T. (2016). New regression model for predicting horizontal forces of single tines using a dummy variable and tine geometric parameters. *LANDTECHNIK*, 71(5), 168–174.

Al-Neama, A., & Herlitzius, T. (2017). Draft forces prediction model for standard single tines by using principles of soil mechanics and soil profile evaluation. *LANDTECHNIK*, 72 (3), 157–164.

Aluko, O., & Seig, D. (2000). An experimental investigation of the characteristics of and conditions for brittle fracture in two-dimensional soil cutting. *Soil & Till. Res.*, (57), 143-157.

Baker, C., & Saxton, K. (2007). *No-tillage Seeding in Conservation Agriculture* (second edition Ausg.). UK: CABI.

Boardman, J., & Poesen, J. (2006). *Soil Erosion in Europe.* England: John Wiley & Sons, Ltd.

Chaudhuri, D. (2001). Performance evaluation of various types of furrow openers on seed drills-a review. *J. of Agric. Eng. Res.*, 79 (2): 125-137.

Conte, O., Levien, R., Debiasi, H., Sturmer, S., Mazurana, M., & Muller, J. (2011). Soil disturbance index as an indicator of seed drill efficiency in no-tillage agrosystems. *Soil and Tillage Res.*, S. 114, 37–42.

Darboux, F., & Huang, C. (2003). An instantaneous-profile laser scanner to measure soil surface microtopography. *Soil Sci. Soc. Am. J.*, 67, 92–99.

Dedousis, A., & Bartzanas, T. (2010). *Soil Engineering.* Berlin: Springer-Verlag Berlin Heidelberg.

Desbiolles, J. M., Godwin, R. J., Kilgour, J., & Blackmore, B. S. (1997). A Novel Approach to the Prediction of Tillage Tool Draught using a Standard tine. *J. agric. Engng Res.*, 66 , 295 – 309.

Desir, F. (1981). *A feld evaluation of the wedge approach to the analysis of soil cutting by narrow blades.* M.Sc. thesis, McGill Univ., Dept. of Agric. Eng., Montreal, Canada.

Dransfield, P., Willatt, S., & Willis, A. (1964). Soil-to-implement reaction experienced with simple tines at various angles of attack. *J. of Agric Engng Res*, 9, 220–224.

Ehrhardt, J., Grisso, R., Kocher, M., Jasa, P., & Schinstock, J. (2001). Using the Veris Electrical Conductivity Cart as a Draft Predictor. *ASAE*, Paper No. 011012.

Elijah, D., & Weber, J. (1971). Soil Failure and Pressure Patterns for Flat Cutting Blades. *Transactions of the ASAE*, 781-785.

Glancey, J. L., & Upadhyaya, S. K. (1995). An improved technique for agricultural implement draught analysis. *Soil and Till. Res.*, 175-182.

Glancey, J. L., Upadhyaya, S. K., Chancellor, W. J., & Rumsey, J. W. (1996). Prediction of agricultural implement draft using an instrumented analog tillage tool. *Soil and Till.Res.*, 37:47-65.

© Springer-Verlag GmbH Germany, part of Springer Nature 2019
A. K. A. Al-Neama, *Evaluation of performance of selected tillage tines regarding quality of work*, Fortschritte Naturstofftechnik, https://doi.org/10.1007/978-3-662-57744-8

Godwin, R. J. (2007). A review of the effect of implement geometry on soil failure and implement forces. *Soil & Tillage Research*, 97, 331–340.

Godwin, R., & O'Dogherty, M. (2007). Integrated soil tillage force prediction models. *Journal of Terramechanics*, (44); 3–14.

Godwin, R., & Spoor, G. (1977). Soil failure with narrow tines. *J. of Agric. Eng. Res.*, 22(4):213-228.

Godwin, R., Spoor, G., & Soomro, M. (1984). The effect of tine arrangement on soil forces and disturbances. *J.Agric. Eng. Res.*, 30:47–56.

Gorjatschkin, W. (1927). *Theory of the plow*. Moscow.

Grisso , R., Yasin , M., & Kocher , M. (1996). Tillage tool forces operating in silty clay loam. *Transactions of the ASAE*, 39,1977–1982.

Grisso, R., & Perumpral, V. (1985). Review of models for predicting performance of narrow tillage tool. *Transactions of the ASAE*, 28 (4): 1062-1067.

Guhr, M. (2015). Experimentelle Untersuchung von Grubberscharen unter Laborbedingungen. Diplomarbeit thesis, TU Dresden, Germany.

Gupta, P., Gupta, C., & Pandey, K. (1989). An analytical model for predicting draft forces on convex-type cutting blades. *Soil and Till. Res.*, 14, 131-144.

Hasimu, A., & Chen, Y. (2014). Soil disturbance and draft force of selected seed openers. *Soil and Till. Res. J.*, 140, 48–54.

Hettiaratchi, D. (1993). The development of a powered low draft tine cultivator. *Soil and Tillage Res.*, S. 28, 159-177.

Hettiaratchi, D., & Reece, A. (1967). Symmetrical three-dimensional soil failure. *J.Terramechanics*, 4(3):45-67.

Hettiaratchi, D., & Reece, A. (1974). The calculation of passive soil resistance. *Geotechnique*, 24: 289-310.

Hettiaratchi, D., Whitney, B., & Reece, A. (1966). The calculation of passive pressure in two-dimensional soil failure. *J.Agric. Eng. Res.*, 11(2):89-107.

Huang, C., & Bradford, J. (1992). Applications of a laser scanner to quantify soil microtopography. *Soil Sci. Soc. Am. J.*, 56, 14–21.

Jester, W., & Klik, A. (2005). Soil surface roughness measurement—methods, applicability, and surface representation. *Catena*, S. 64, 174–192.

Kasisira, L. (2004). *Force modelling and energy optimization for subsoilers in tandem*. PhD thesis, University of Pretoria, South Africa.

Koolen, A. (1977). *Soil loosening processes in tillage analysis, systematics and predictability*. Wageningen, The Netherlands: Tillage Laboratory, Agricultural University.

Koolen, A., & Kuipers, H. (1983). *Agricultural Soil Mechanics*. Berlin Heidelberg: Springer-Verlag.

Liu, J., & Kushwaha, R. L. (June 26 - 29, 2005). Study of soil profile of a single sweep tool. *CSAE/SCGR Meeting* (pp. Paper No. 05-060). Canada,Winnipeg, Manitoba: CSAE/SCGR.

Liu, J., Chen, Y., Lobb, D., & Kushwaha, R. (2007). Soil-straw-tillage tool interaction: Field and soil bin study. *Canadian Biosystems Engineering*, 49, 2.1-2.6.

Luth, H., & Wismer, R. (1971). Performance of plane soil cutting blades in sand. *Transactions of the ASAE*, 14 (2), 255-259.

Makanga, J., Salokhe, V., & Gee-Clough, D. (1996). Effect of tine rake angle and aspect ratio on soil failure patterns in dry loam soil. *J. of Terramechanics*, 33, 233–252.

Manuwa, S. (2009). Performance evaluation of tillage tines operating under different depths in a sandy clay loam soil. *Soil & Tillage Research J.*, (103), 399–405.

Manuwa, S. I., Ademosun, O. C., & Adesina, A. (2012). Regression Equations for Predicting the Effect of Tine Width on Draught and Soil Translocation in Moderately Fine Textured Soil. *Journal of Environmental Science and Engineering*, 820-825.

Manuwa, S., & Ademosum, O. (2007). Draught and soil disturbance of model tillage tine under varying soil parameters. *CIGR E journal*, 1-18.

Manuwa, S., & Ogunlami, M. (2010). Soil–tool interaction modeling of parameters of soil profile produced by tillage tools. *J. Eng. Appl. Sci.*, 5 (2), 91–95.

Martinez-Agirre, A., Álvarez-Mozos, J., & Giménez, R. (2016). Evaluation of surface roughness parameters in agricultural soils with different tillage conditions using a laser profile meter. *Soil and Till.e Re.*, 161, 19–30.

McKyes, E. (1985). *Soil Cutting and Tillage.* New York: Elsevier publishing.

McKyes, E. (1989). *Agricultural engineering soil mechanics.* New York: Elsevier publishing.

McKyes, E., & Ali, O. (1977). The cutting of soil by a narrow blade. *J. of Terramechanics*, 14(2):43-58.

McKyes, E., & Desir, F. (1984). Prediction and field measurements of tillage tool draft forces and efficiency in cohesive soils. *Soil & Tillage Res.*, 4, 459-470.

Mckyes, E., & Maswaure, J. (1997). Effect of design parameters of flat tillage tools on loosening of a clay soil. *Soil and Tillage Res.*, 43,195-204.

Moeenifar, A., Kalantari, D., & Seyedi, S. (2013). Application of dimensional analysis in determination of traction force acting on a narrowblade. *International Journal of Agriculture and Crop Sciences (IJACS)*, 5 (9), 1034-1039.

Moeenifar, A., Mousavi-Seyedi, S., & Kalantari, D. (2014). Influence of tillage depth, penetration angle and forward speed on the soil/thin-blade interaction force. *Agric. Eng. International: CIGR*, 16 (1), 69-74.

O'Callaghan, J., & Farrelly, K. (1964). Cleavage of soil by tined implements. *J. agric. Engng. Res.*, 9(3): 259–270.

Oelze, M., Sabatier, J., & Raspet, R. (2003). Roughness measurements of soil surfaces by acoustic backscatter. *Soil Sci. Soc. Am. J.*, 67 (1), 241–250.

Onwualu, A., & Watts, K. (1998). Draught and vertical forces obtained from dynamic soil cuttting by plane tillage tools. *Soil and Till. Res.*, 48(4):239-253.

Osman, M. (1964). The mechanics of soil cutting blades. *Agric. Eng. Res.*, 9(4):313-328.

Payne, P. C. (1956). The relationship between mechanical properties of soil and the performance of simple cultivation implements. *J.Agric. Eng. Res.*, 1 (1): 23–50.

Payne, P., & Tanner, D. (1959). The relationship between rake angle and the performance of simple cultivation implements. *J. of Agric. Eng. Res.*, 4 (4), 312–325.

Perumpral, J., Grisso, R., & Desai, C. (1983). A soil tool model based on limit equilibrium analysis. *Transactions of the ASAE*, 26(4):991-995.

Rahman, S., & Chen, Y. (2001). Laboratory investigation of cutting forces and soil disturbance resulting from different manure incorporation tools in a loamysand soil. *Soil and Tillage Res.*, 85, 19-29.

Rahman, S., Chen, Y., & Lobb, D. (2005). Soil Movement resulting from Sweep Type Liquid Manure Injection Tools. *Biosystems Engineering J.*, 91 (3): 379-392.

Reece, A. (1965). The fundamental equation of earth-moving mechanics. *In Symposium on Earth-Moving Machinery, Proceedings ofthe Institution of Mechanical Engineers*, 179: 8-14.

Riegler, T., Rechberger, C., Handler, F., & Prankl, H. (2014). Image processing system for evaluation of tillage quality. *Landtechnik*, 69 (3), 125–131.

Rosa, U., & Wulfsohn, D. (2008). Soil bin monorail for high-speed testing of narrow tillage tools. *Biosystems Engineering J.*, (99), 444 – 454.

Rowe, R., & Barnes, K. (1961). Influence of speed on elements of draft of a tillage tool. *Transactions of the ASAE*, 4, 55-57.

Sahu, R., & Raheman, H. (2006). Draught Prediction of Agricultural Implements using Reference Tillage Tools in Sandy Clay Loam Soil. *Biosystems Engineering, SW-Soil and Water*, 88 (2): 275–284.

Saleh, A. (1993). Soil roughness measurement: chain method. *J. Soil Water Conserv.*, 48 (6), 527–529.

Schuring, D., & Emori, R. (1964). Soil deforming processes and dimensional analysis. *SAE New York, paper No. 897 C.*

Sharifat, K., & Kushwaha, R. (2000). Modelling soil movement by tillage tools. *Canadian Agri. Eng. J.*, 42 (4):165-172.

Shen, J., & Kushwaha, R. (1998). *Soil-Machine Interactions: A Finite Element perspective.* USA: Marcel Dekker, Inc., New York. Basel. Hong Kong.

Shinde, G. U., Badgujar, P. D., & Kajale, S. R. (2011). Experimental Analysis of Tillage Tool Shovel Geometry on Soil Disruption by Speed and depth of operation. (S. 65-70). Singapore: IACSIT Press, Singapore.

Siemens, J., Weber , J., & Thornburn , T. (1965). Mechanics of soil as influenced by model tillage tools. *Transactions of the ASAE*, 8, 1–7.

Söhne, W. (1956). Einige Grundlagen für eine landtechnische Bodenmechanik. *Grundlagen der landtechnik*, Vol. 7, 11-27.

Sokolovski, V. (1965). *Statics of soil media.* Oxford, London: Pergamon Press.

Solhjou, A. (2013). *Study into the mechanics of soil translocation with narrow point openers.* PhD thesis, Barbara Hardy Institute, University of South Australia, Australia.

Solhjou, A., Desbiolles, J., & Fielke, J. (2013). Soil translocation by narrow openers with various blade face geometries. *Biosystems Engineering*, S. 114 (3), 2 5 9 - 2 6 6.

Soucek, R., & Pippig, G. (1990). *Maschinen und Geräte für Bodenbearbeitung Düngung und Aussaat.* Germany: Technik GmbH Berlin.

Spoor, G., & Godwin, R. (1978). An Experimental Investigation into the Deep Loosening of Soil by Rigid Tines. *agric. Eng. Res.*, (23), 243-258.

Sprinkle, L., Langston, T., Weber, J., & Sharon, N. (1970). A similitude study with ststic and dynamic parameters in an artificial soil. *Transactions of the ASAE*, 13, 580-586.

Stafford, J. (1979). The Performance of a Rigid Tine in Relation to Soil Properties and Speed. *J. agric. Engng. Res.*, (24), 41-56.

Stafford, J. (1981). An application of critical state soil mechanics: the performance of rigid tines. *J. agric. Eng. Res.*, 26 (5): 387- 401.

Stafford, J. (1984). Force prediction models for brittle and flow failure of soil by draught tillage tools. *J. agric. Engng Res.*, 29, 51-60.

Stout, B. A., & Cheze, B. (1999). *CIGR Handbook of Agricultural Engineering Volume III Plant Production Engineering.* United States of America: ASAE.

Swick , W., & Perumpral , J. (1988). A model for predicting soil-tool interaction. *J.of Terramechanics*, 25, 43–56.

Terzaghi, K. (1943). *Theoretical Soil Mechanics.* New York: John Wiley and Sons Inc.

Upadhyaya, S. K., Williams, T. H., Kemble, L. J., & Collins, N. E. (1984). Energy requirements for chiseling in coastal plain soils. *Transactions of the ASAE*, 27(6):1643-1649.

Van-Camp, L., Bujarrabal, B., Gentile, A. R., Jones, R. J., Montanarella, L., Olazabal, C., & Selvaradjou, S.-K. (2004). *Reports of the Technical Working Groups Established under the Thematic.* Luxembourg: Office for Official Publications of the European Communities.

Wheeler, P. N., & Godwin, R. J. (1996). Soil Dynamics of Single and Multiple Tines at Speeds up to 20 km / h. *J. agric. Engng Res.*, 63 , 243 – 250.

Willatt, S. T., & Willis, A. H. (1965). A Study of the Trough Formed by the Passage of Tines through Soil. *J Agric. Eng. Res.*, 10, 1 - 4.

Zeng, D., & Yao, Y. (1992). A dynamic model for soil cutting by blade and tine. *J. Terramechanics*, 29, 317–327.

List of Figures

© Springer-Verlag GmbH Germany, part of Springer Nature 2019
A. K. A. Al-Neama, *Evaluation of performance of selected tillage tines regarding quality of work*, Fortschritte Naturstofftechnik, https://doi.org/10.1007/978-3-662-57744-8

List of Tables

Printed in the United States
By Bookmasters